Sound Studies Series

Series edited by Holger Schulze Volume 5

Sound Studies

The book series Sound Studies presents research results, studies and essays on a rather new yet well-known field of research:

How do human beings and animals and things live together with all the sounds and noises, tones and signals of their times? How do they shape and design sounds – and how do they act through sounds and explore their world, even in foreign or in maybe only superficially known cultures?

The research field of Sound Studies is transdiciplinary and transmethodologically by nature: the publications of this book series therefore present artistic and design concepts from fields such as sound art, composition, performance art, conceptual art and popular culture as well as articles from disciplines such as cultural studies, communication studies, ethnography and cultural anthropology, music studies, art history and literary studies. Artistic research as a whole is therefore an important approach in the field of sound studies.

The Sound Studies book series intends to open up a discourse in, on and about sound – across the boundaries of academic disciplines and methods of research and artistic invention: a speaking about sound beyond the hitherto alltoo well-known academic discourse.

Series editor: Holger Schulze

A publication of the Sound Studies Lab at the Humboldt-Universität zu Berlin.

Bibliographic information published by the Deutsche Nationalbibliothek
the Deutsche Nationalbibliothek lists this publication in the Deutsche Nationalbibliografie; detailed bibliographic data are available in the Internet at http://dnb.d-nb.de

Cover image: Moniek Wegdam, Dam, Amsterdam, February 13, 2012
Editorial Staff: Georg Spehr, Berlin
Cover layout & typeset: Christina Giakoumelou, Athens
Printed by Majuskel Medienproduktion GmbH, Wetzlar
ISBN 978-3-8376-2179-2

Karin Bijsterveld (ed.)

Soundscapes of the Urban Past.
Staged Sound as Mediated Cultural Heritage

[transcript]

Contents

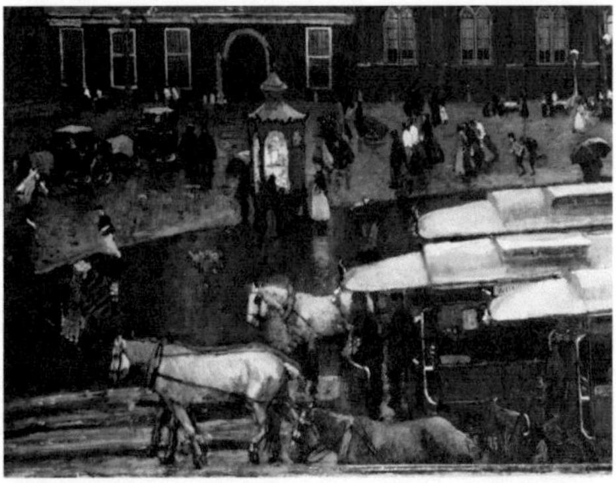

Introduction

Karin Bijsterveld

1. »Take a Closer Listen«

It happened when his Ipod was out of batteries. Lying in a park in Geneva, on a midsummer day, with his headphones still on, he suddenly heard the sounds of the park as if played on his Ipod: people talking, a dog barking, a truck passing, kids laughing, birds singing. The idea for a book was born. Artist Rutger Zuyderfelt invited people to »take a closer listen« to the sounds that normally go unnoticed to them, and asked them to think of their »favourite sound«. By the end of 2009, he had gathered fifty-two detailed descriptions of sounds, as well as a few accompanying stories. [1]

Many of the sounds submitted were those of nature: the »flapping of birds' wings«, or »wind blowing through trees«. Some were sounds of nature audible in enclosed, man-made spaces, like »rain falling on a tent's roof« and »snow lightly hitting a glass window«. Other stories just referred to the sounds of people talking or things moving: »neighbours' muffled conversation«, »umbrellas opening and closing« or a »far-off foghorn from a ship at sea«. Still others, however, referred to the sounds of crowded, urban environments such as »rumbling train engines«, »constant traffic on a distant road«, a »busy playground several yards away«, or »echoes in a tube station«. And some seemed to be associated with joyful expectation, such as »mail falling on a doormat«, »a new pack of coffee being opened« and »a book page being turned«. [2]

Some seven decades before, in the mid-1930s, residents of the city of Groningen, The Netherlands were also invited to take a closer listen. The organization involved was a

1 Zuyderfelt 2009, 3, 5, 7.

2 Zuyderfelt 2009, 10-73.

local noise abatement committee, one similar to those established in many other Western towns and cities in the first decades of the twentieth century, including New York, London and Berlin, to mention a few metropolitan examples.[3] The Groningen committee publicly announced its plan to take action against unnecessary noise, and asked local residents to provide information on the sounds they considered annoying. This triggered a response of over one hundred letters.[4] Most of the letter-writers complained about »street noise«, notably the sounds of the horns and engines of cars, trucks and motor bikes, but also the noises of municipal street cleaning services, carts, trams, trolley buses, trains, and people loading and unloading cargo. Others felt bothered by the sounds of local tower carillons, factories and steam whistles, street vendors, public loudspeakers, soldiers doing firing exercises, children, dogs, cats and cocks. Quite a few also had concerns about the loudness of their neighbours' radio.

The complainants often stressed the incessant and »piercing« character of the unwanted sounds, their sheer multitude, as well as their untimely nature: a »screaming army« of flower vendors, rag-dealers and iron mongers passing through the street »day and night«. Each morning, a family woke up far too early due to milk factory workers »throwing with« their milk »tins« and »lids«. Other commercial activities, such as the outdoor second-hand car market, would start as early as in the middle of the night, with »slamming car doors« and »endless« starting up of engines »in vain«, for not all cars offered for sale there were »of the modern kind«. Motorized traffic contributed to the hubbub, but so did the numerous carts delivering milk or vegetables, which had »no rubber tires« while the street pavement was rather »uneven«. In their letters, the writers also voiced medical concerns (»I have been seeing the doctor for over a year now and suffer a lot from nervousness«) or claimed that »the sick« in their family could not stand the noise anymore, hoping that the committee's members could do something about it. Furthermore, the noise of motor bikes, driven up and down the road by the neighbourhood street kids (sons of the »knife-grinders and oil millers« and a few »pimps«) would cause houses to »shake«, while the drone of electric engines felt, as one letter put it, »as if my brain begins to tremble«. Hospital patients, one writer knew, suffered from sleep disturbance caused by a local carillon, as was true of »strangers visiting our town for just a few days«. The city was overcrowded with »twitter bicycles«,

3 Bijsterveld 2008.

4 Archives Anti-lawaai comité Groningen (ALCG, Noise Abatement Committee Groningen), File »Brieven op Persberichten« (Letters in response to Press Reports), 1935-1937; courtesy Hero Wit, now stored at Maastricht University, Faculty of Arts & Social Sciences. The committee itself numbered the letters to which we refer.

»car din« and an outright »chaos of sounds«. One complainant regretted that those from the nearby countryside (»*buitenmenschen*«) would often go into Groningen very early, at a time when the urban residents (»*stadjers*«) were still sound asleep, »as they go to bed later«. Another contended that it was usually »lower-class lads and girls« who felt a need to scream and yell.[5]

We do not present these examples to suggest that urban dwellers' appreciation of urban sound used to be more negative in the past than it is today. It might have been just as easy to find negative stories about present-day urban sounds, or, conversely, positive ones about such sounds in the past. Rather, our interest is in how these stories are staged. Intriguingly, many of the stories about favourite sounds generally speak of sounds that are hardly discernible, such as muffled sounds or lightly tapping ones. These sounds also tend to be produced at a certain distance from the listener: far-off, several yards away, amounting to echoes rather than the real thing. We classify them as typical instances of the »comforting sound« topos, a description or staging of sound which foregrounds people's experience of safety, a calm atmosphere, a sense of security. We will explain this topos in more detail below. In contrast, most of the Groningen letters contained examples of »intrusive sound«, of discourse on violent sounds that, often repeatedly, intrude people's private spaces without being appreciated by them. In their narrative structure and style, the letters highlight the inescapability of noise by enumerating the many different sources of sound and mentioning their recurrence; or by emphasizing the weakness of the complainants or others suffering: hospital patients, the sick, neurotics or children. An alternative rhetorical strategy adopted is to present the noise-makers as people of a lesser kind (boys, lower-class youths, or pimps) than the complainants – all honourable citizens by implication.

This book addresses all these various ways of »staging sound«. It focuses, however, not only on how sounds are captured in text – the textualization of sound – but also on the staging and dramatization of sound in radio and film. And it is not just about all the sounds one can imagine, but about the sounds of the *city* and how these urban sounds have been staged in texts, radio plays and film productions created in the long twentieth century.

5 All quotes in this paragraph are from the ALCG-letters, respectively: 99, 102, 103, 107, 105, 101, 103, 107, 80, 100, 106, 64, 104, 102, 80, 83 and 66.

2. Media and the Staging of Sound. Textualization and Dramatization

What is the use of our concern with the various ways urban sounds have been »staged«? The Groningen examples quoted above underscore that sound frequently figures as a deeply contested phenomenon, most notably in urban, densely populated areas. What is music to the ears of some residents may be unwanted sound, and thus noise, to the ears of their neighbours. Such clashes over urban sounds do not only touch on their individual meanings to those who hear them, but also express ideas about what the character of the city should be like, and what is allowed to be audible or not. Should we, for instance, conceive of cities as dynamic domains whose charm partly, if not largely, derives from the complex symphony of sounds they generate on a daily basis, or should we rather conceive of them as places of important intellectual work where all residents are entitled to find the peace and tranquillity needed to work diligently and productively?

Past sounds and the perception of these sounds by historical actors can also inform us about the changing character and identity of cities. We usually do not have direct access to these past sounds, however. There are very few recordings of everyday Western life before 1900, in part because early anthropologists focused on making recordings of *non*-Western societies.[6] And for the years up until World War II, most recordings of everyday sonic environments were in fact made for radio plays and films. Our knowledge of past soundscapes, transient and intangible as they are, is therefore largely dependent on historical texts in which people described what they heard and what these sounds meant to them. At the same time, however, our *imagination* of such soundscapes has been nourished by the soundtracks the makers of radio plays and films created for their productions. And it is this *mediated* cultural heritage of sound that presents us with a unique chance to study the dramatization of urban sound over time, and thus to understand the varying and changing representations of urban identities. This has been the objective of the research project this volume draws on: »Soundscapes of the Urban Past: Staged Sound as Mediated Cultural Heritage«.

Methodologically, we – the members of the Soundscapes of the Urban Past-team Jasper Aalbers, Karin Bijsterveld, Andreas Fickers and Annelies Jacobs – started from the view that representations of sound, both in historical text and in radio play and fiction film, always imply a particular dramatization of sound. It has been common for historical actors to articulate their perception of sounds only when particular sounds *moved* them, in either a positive or a negative sense. In making their responses to sound ex-

6 Brady 1999.

plicit, and thereby rhetorically strengthening their position, they employed particular *repertoires of dramatizing sound*. But not only historical actors did so. Makers of radio plays and fiction film have similarly developed repertoires, and at times genre-specific repertoires, for dramatizing sound in order to articulate their take on the protagonists of their stories or the settings in which they situated these stories. We have therefore treated documents referring to putative historical realities on the same plane as media expressing fictional realities, by focusing on the repertoires of dramatizing sound they had in common, while also being open to the genre-specific differences in such repertoires. Among the repertoires we started with were *auditory topoi* such as the intrusive sound, the sensational sound, the comforting sound, and the sinister sound; we already mentioned two of these frequently used forms of staging sound above.[7] We used these auditory topoi as a key tool for analyzing the dramatization of sound in historical texts, radio plays and fiction films, and the same applies to several other *narrative* strategies for capturing sonic experiences.

In addition, we examined the dramatization of sounds in terms of *keynote sounds*, *sound marks* and *sonic icons* featuring in these sources. The notion of »keynote sounds« refers to sounds that make up the background sound of a sonic environment, such as traffic sound in the modern city. In music, it finds its analogies in the keynote of a composition as well as in the basso ostinato, a repeated bass line. »Sound marks« refer to the sounds that stand out in a particular environment, and are considered typical for a specific location, such as the bell of London's Big Ben or the phrase »mind the gap« in the London Underground.[8] Over time, particular sounds can become *iconic* for particular locations through the conventional ways in which they are deployed in written narrative, in film and on the radio, as well as through subsequent inter-textual references. A good example is the brief accordion tune that for many has come to evoke the city of Paris. In theory, such *sonic icons* may not bear any relationship, or no relationship anymore, with the actual location – the iconicity is merely a convention. Often, however, particular sound marks from the past somehow persist and become iconic over time. In addition, particular sounds may acquire a narrative iconicity. For example, the sounds of arriving coaches and cars, of door bells ringing, of doorsteps and wheels crunching the gravel have come to signify shifts in plots and scenes in a canonized manner.[9] A similar example is the routine representation of landing airplanes in films through the sound of the screeching of wheels touching the tarmac.

7 Bijsterveld 2008.

8 Schafer 1994 [1977].

9 Krug 2003, Thomsen / Schneider 1985.

The artificial iconic character of this sound is even more evident to those who know that in reality we cannot hear the wheels, owing to the masking effect of the roaring engines. But sounds may also perform narrative roles in less conventional and more surprising ways: they create additional drama, suspense or identification marks.[10] A seminal example is the sonic »close up« Alfred Hitchcock employed in *Blackmail* (1929). He amplified the word »knife« among incomprehensible babble in order to highlight the significance of this word to the protagonist. This signifier, in turn, elicited the connotations of strength, significance, and aggressive intrusion that loud sounds have often had in Western culture.[11] A focus on the iconization of sound in film – as in radio plays and historical text – thus also allows for an analysis of the particular ways in which sounds have been loaded with meaning.

As concepts, »keynote sounds« and »sound marks« were coined by the composer and environmentalist Raymond Murray Schafer in the 1970s. His publications opened up the historiography of sound by documenting and reflecting on changes in the Western soundscape, or sonic environment, since the Industrial Revolution.[12] (For a detailed discussion of the notion of »soundscape«, see Jonathan Sterne's chapter, this volume.) Initially, historians of sound started out in a similar vein: they catalogued all the sounds citizens could hear at particular moments in time.[13] Historian Alain Corbin stressed, however, that it is at least as interesting to examine how citizens' past habitus conditioned their *ways* of listening. Which sounds did they listen to most attentively, and why? Which sounds seemed simply below their thresholds of perception? And what were, in terms of auditory and other sensory impressions, the configurations of the tolerable and intolerable in past societies?[14] Corbin's work on bells in the nineteenth-century French countryside has revealed how bells not only structured the villagers' days and mediated news in ways we would not able to understand today, but also how they contributed to people's spatial orientation and expressed the symbolic power of towns.[15] More recent publications on sound give ample attention to such cultural meanings of sound and shifting modes of listening.[16] Several even focus on the chang-

10 Pisters 2007, Lastra 2000.

11 Bijsterveld 2001.

12 Schafer 1994 [1977].

13 Thuillier 1977.

14 Corbin 1995.

15 Corbin 1999 [1994].

16 See, for instance: Thompson 2002, Bull / Back 2003, Bijsterveld 2003, Rath 2003, Sterne 2003, Smith 2004, Cockayne 2007, Parr 2010, Morat et al. 2011, Müller 2011, Schwartz 2011, Birdsall 2012, Pinch / Bijsterveld 2012.

ing »sensescapes« of *cities* in particular, even though sound is only one of the sensorial experiences under study.[17]

From these studies we know that many conflicts about sound have involved issues of power and the right to dominate some environment with specific sounds or, conversely, to free a setting from such sounds. The urban campaigns in the late 1800s by the intellectual elite against street musicians, for instance, were an expression not only of noise abatement, but also of annoyance about the nearby presence of people lower in class and of foreign origin, and as such they were related to the rise of a new professional class of urban writers, journalists and professors working at home.[18] Similarly contested sounds were the rhythmical sounds of mechanical industry and the chaos of urban traffic. If some interpreted these as sounds of progress, others saw them as vulgar attacks on distinguished minds.[19]

While many of these studies address how urban dwellers have regulated these sounds and even adapted to them over time, the sonic icons used in radio plays and films may have canonized alternative representations of the modern city, for instance by stressing the chaotic bombardment of the ear over the comforting aspects of urban sound. To develop a more layered argument, then, we compare the dramatization of urban sound in historical documents with the dramatization of urban sound in radio plays and films. This allows us to capture differences and similarities in the representation of urban identities through the staging of sound *across* different media. To prevent overgeneralization in our conclusions about the urban, however, we have confined our exploration of the staging of urban sound to *three cities* in particular: Amsterdam, Berlin and London. These capital cities inject, as Jill Steward and Alexander Cowan have said about such cities, »into the commercial activity common to all cities a strong administrative element as well as all the richness and theatre of public rituals.«[20] At the same time Amsterdam interestingly differs from Berlin and London in not having their metropolitan size.

Another self-imposed restriction is temporal: we only cover the long twentieth century, years of intense motorization and »multi-culturalization«, although our sources on radio plays and sound films, of course, only go back as far as the 1920s and 1930s, the decades of their rise. Yet, while all members of the Soundscapes project team had a focus of their own – Annelies Jacobs studied the dramatizations of sound in historical text

17 Adams / Guy 2007; Cowan / Steward 2007, Diaconu/Heuberger/Mateus-Berr/Vosicky 2011. See for an overview of relevant literature also Howes 2011, 66.

18 Picker 2003.

19 Smith 2001, Bijsterveld 2001.

20 Steward/ Cowan 2007, 9.

on Amsterdam; Andreas Fickers unravelled the representations of Berlin, Amsterdam and London soundscapes in radio plays; and Jasper Aalbers examined the sound of fiction films with plots situated in these three cities[21] – this book centers on the *synchronic* rather than diachronic aspects of their analysis. It explicitly *compares* the staging of urban sound *across* our cultural heritage of text, radio and film: which repertoires of dramatizing urban sound did these media have in common or not, and how did these contribute to the collective of stories we tell about cities, and thus to urban identities? This media-comparative approach to sound is rather new, and therefore our undertaking has a highly explorative character.

Auditory Topoi, Acoustic Profiling and Audiographs

We have touched upon the notion of auditory topoi several times, but have not yet explained it in more detail. One of us has distinguished four auditory topoi – intrusive sound, sensational sound, comforting sound and sinister sound – in a study of public debates on the problem of noise in Western societies in the long twentieth century, *Mechanical Sound*.[22] This typology of auditory topoi, inspired in part by work of Michael Cowan[23] and Philipp Schweighauser[24] on the textualization of sound and noise in poetry and literature, originally drew on an analysis of quotes in Western, largely realist novels published between 1875 and 1975 on urban, industrial or mechanical sounds, often involving sounds that resulted from modern technologies. It was possible to distinguish the four ideal-types by classifying these quotes in terms of the *quantity*

21 Annelies Jacobs focused on four episodes in the history of Amsterdam, 1875-1895, 1918-1940, 1940-1945, and 2005-2010, periods known to display important transitions in urban sound. Newspapers, travel-stories, diaries, city council reports and police archives have been her most important sources, see Jacobs 2013 forthcoming. Andreas Fickers studied three different types of radio plays produced since the late 1920s: radio plays as »theater of the blind«, adaptations of literary works for radio, and experimental radio plays such as »sound portraits«, »sound montage«, »sound symphonies« and »acoustical films« (see Döhl 1985, 40-59, and Cory 1994, 337). Jasper Aalbers studied a selection of 95 films produced in the 1930s, 1950s, 1970s and 1990s, and set in Amsterdam, London or Berlin. See Aalbers 2013 forthcoming.

22 Bijsterveld 2008.

23 Cowan 2006.

24 Schweighauser 2006.

of sound-sources described, the *distance* between the protagonist or narrator and the sound, the *direction* of the sound, and its *rhythm*.

	Evaluation of sound			
	Negative		**Positive**	
	Intrusive	*Sinister*	*Sensational*	*Comforting*
Quantity	Multitude	Single	Multitude	Single
Distance	Close	Diverse	Diverse	Far
Direction	→ subject	Unclear	→ and ←	Unspecified
Rhythm	Irregular or Unpredictable	Unspecified	Regular	Regular or Unspecified

Auditory Topoi of Technology (Western novels, 1875-1975)

Intrusive sound is expressed as a multitude of different sounds or a series of recurrent sounds. They invade or threaten the existence of a vulnerable or fragile quality, such as nature, harmony or one's heart, mind, body or security. The noise frightens the protagonist and seems to move ever closer, if not penetrating him or her. *Sensational* sound is the positive counterpoint of intrusive sound. Like intrusive sound, sensational sound refers to a multitude of sounds, such as the crowds of the city or traffic. Unlike intrusive sound, however, the sources of sensational sound can be felt both close and rather far from the narrator or protagonist, and, in all cases, they fill the environment and surround or »lift« the subject in an enthusing way. *Comforting* sound, like sensational sound, has a positive ring. Yet unlike sensational sound, a comforting one is a single source of sound, or a single type of source. It is often a sound heard at a certain distance, usually from an unspecified direction, but even when described as nearby it does not infringe on the narrator's or protagonist's subjective experience. In all cases the shelter, security and harmony of the narrator or protagonist's direct environment (the bed, the house) are highlighted. A *sinister* sound, finally, is usually a single sound within a more or less silent environment. It functions as an ominous sign of what is ahead. Its distance may be close or indeterminate, and its direction is often unclear. It is the sound of suspense, like a creaky door in a haunted house.

The variety of sounds categorized as belonging to one topos or another suggests that such stereotypical expressions of sound in literature are more or less independent from the source of the sound. The sinister sound, for instance, refers to both human and non-human or mechanical and non-mechanical sources of sound. At the very same time, however, the quotes analyzed highlight particular aspects of mechanical sound. In the topoi of both sensational and intrusive sound, it is the multitude, movement and continuity of the sounds of traffic and machinery that increase the tension. Although the single scream of a tramcar or whip-crack can be striking, it is the recurring character of the phenomenon and the mix of all sounds together that heighten the drama in particular. In the case of sensational sound, it is the collective and ongoing rhythm that contributes to the textual staging of their positive effects. In the case of intrusive sound, it is often the irregularity, chaos, and unpredictability of sounds – the opposite of cadence and rhythm – that contributes to the dramatic performance.

The presentation of distance and direction do additional »dramatic work«. The continuity of comforting sounds creates a soothing hum, but their distance and unspecified direction ensure their innocence. In contrast, the movement of sounds towards the narrator helps to stylize sound as intrusive noise, whereas lack of knowledge of the exact location and source of sound contributes to its dramatization as being a malevolent or creepy sound. Mechanical repetition, however, is responsible for the inexorability *and* the untiring animation of sound, and referring to the continuous character of mechanical sounds helps conjuring up the soothing spirit of the comforting sound, the dynamics of the sensational sound, and the inescapability of the intrusive sound. Given the »mechanized« character of the twentieth-century city, it is likely that the auditory topoi of the intrusive, sensational, comforting and sinister sound are also detectable in the representation of urban soundscapes in historical texts, radio plays and films.

We have gradually acknowledged, however, that our research in »Soundscapes of the Urban Past« should be equally open to what the representation of voices in the urban context can do. As Greg Goodale shows in his recent book *Sonic Persuasion. Reading Sound in the Recorded Age*, the use of dialect on early twentieth-century American sound recordings of dialogue often signified urban immigrants, and played out elite pronunciation against lower-class dialect and immigrant slang.[25] And Philip Schweighauser has explained how American writers in late nineteenth century and after recorded such new immigrant voices in their novels. Often these writers emphasized the »obscurity of foreign-sounding speech«, as if immigrant voices produced »nothing but unintelligible noise«, or depicted their voices as excessively loud, hoarse and croaking, ways of »acoustic profiling« along the lines of class and race.[26] In case the writers furnished a particular protagonist with a consistent acoustic profile, Schweighauser even speaks of an »audio-

25 Goodale 2010, 16-26.

26 Schweighauser 2006, 62, 65.

graph«, a characterization technique that endows fictional bodies with a set of distinctive acoustic properties designed to position characters with regard to the ensemble of social facts and practices that constitute the fictional world they inhabit. These acoustic properties may range from characters' accents, dialects, or intonation patterns to the sounds produced by their laughter, snoring, or the acoustic impact of their footsteps. The positioning accomplished via audiographs may involve value judgments on the part of other characters, narrators, and implied authors as well as implied and empirical readers.[27]

So on top of the conceptual and methodological equipment we started with – auditory topoi, keynote sounds, sound marks and sonic icons – we took audiographs and acoustic profiling of language and voice on board in order to study the staging of urban sound. In our next chapter, we will explain which narrative tools were additionally important to understand how writers and makers of radio and film »captured« the urban sonic environment. Finally, we had a special interest in films that had their counterpart in radio plays and novels, such as *Berlin Alexanderplatz*, as this facilitated our comparison across different media. Our chapter entitled »Sounds Familiar« will discuss the issues of »intermediality« associated with such re-use of narrations.[28]

4. An Ears-on Experience and the Organization of this Book

As explained above, our interest in the staging of urban sound in text, radio and film has been driven both by the idea that such cultural heritage is a significant entrance into *representations of the urban past*, and that such representations, in turn, are markers of cultural shifts and variations in *urban identities*. A third motivation, not mentioned yet, has been to inspire the employment of such cultural heritage in history *museums* aiming to inform their audiences on the changing meanings of sound in the urban context. Our project and book, we hope, may enrich the ways in which museums present, and again stage, cultural and urban history.

In recent years, museums have fully embraced multimedia in order to attract new audiences. Exhibitions, however, are still predominantly visual in character, even though this book's next chapter will start with examples of auditory exhibitions. It has been one of the aims of our project to create an installation at the Amsterdam Museum[29] which, in

27 Schweighauser 2006, 71.

28 Rajewski 2005.

29 Formerly: Amsterdam Historical Museum.

an interactive manner, presents our knowledge of changing urban soundscapes and the ways in which they have been perceived. Such an exhibition can also literally speak to audiences with visual impairments, a growing group of people in the graying West. The installation entails a reflexive simulation of past soundscapes of Amsterdam that will be presented in a closed room, with pictures serving as decoration for sounds instead of the other way around. We will make the simulation with help of *Virtual Soundscapes™ Technology (VST)*. VST is a digital tool developed by the American company Harris, Miller, Miller & Hansen Inc. for simulating changes in sonic environments. Normally, the tool enables citizens to listen to virtual sonic environments such as a highway lane, and to the effect of interventions – such as noise barriers – on these soundscapes. We aim to apply VST in such a manner that it cannot only simulate *future* soundscapes, but also *past* soundscapes. Doing so requires the recording of obsolete technologies' sounds, and asks for extensive historical research into past conditions and cultural contexts. In contrast to recordings available in sound archives, the tool enables the interactive re-enactment of changes in soundscapes by gradually adding and removing layers of sound, and the reconstruction of masking, shielding and reverberation effects. Our installation will also be a *reflexive* one, by which we mean that we do not merely provide the museum visitors with sounds, but also with information about how historical actors perceived and gave meaning to these sounds.

It is by no means self-evident, however, how such information can be best and most effectively offered to museum visitors. In the Amsterdam case, we have developed, in addition to the virtual soundscapes, three radio play-like storylines for three episodes in the history of Amsterdam – the late nineteenth century, the interwar years and World War II – that convey how Amsterdam residents gave meaning to their changing sonic environments. At the symposium *Staging Sound in the Museum: The Ears-on Experience of History* which took place at the Institute for Image and Sound, Hilversum, the Netherlands on February 3, 2012, we discussed many other options, several of which will be discussed in our next chapter, »Shifting Sounds«. The main thrust of that chapter is, however, to offer museum curators inspiration for presenting auditory histories by showing how authors of texts and makers of radio plays and films used particular strategies for capturing and articulating such soundscapes. In doing so, »Shifting Sounds« both aims to contribute to a comparative study of staging urban sound across different media, *and* to be a source of inspiration to the museum world.

This book has been organized around three major chapters: »Shifting Sounds«, »Sounds Familiar« and »Sonic Artefacts«. The first two were written by the core members of the Soundscapes project, the first with Karin Bijsterveld and the second with Andreas Fickers as lead author, and both chapters have a comparative ambition. While »Shifting Sounds« unravels narrative tools that make urban auditory experiences *tangible* for a few moments, »Sounds Familiar« aims to study intermediality through the lens of sound by focusing on and comparing the staging of urban sound in the novel, radio play, film and television series *Berlin Alexanderplatz*. Its perhaps surprising argument reveals that the richest expression of urban sonic experiences is found in the novel, rather than in the radio play, film and television adaptations, despite their reliance on actual sounds. We asked Carolyn Birdsall to contribute the third major chapter, »Sonic Artefacts«, because we believed it to be worthwhile to study the staging of urban sounds not only in radio fiction, but also in radio documentary. Birdsall zooms in on the reality codes and thus the sonic *artefacts* of early German radio documentary. Just like the other key chapters, this chapter has a media-comparative element. It does not only show how conventions from visual documentary informed radio documentary, but also how visually based technique such as *bird's eye perspective* found its auditory counterpart through a particular use of the microphone in radio.

After the three main chapters were completed, we asked several scholars from Sound and Media Studies to write *expansions* to these chapters. Historian of the Senses Mark M. Smith, Film Studies scholar Patricia Pisters, and Radio studies experts Evi Karanathasopoulou and Andrew Crisell responded to »Shifting Sounds«, »Sounds Familiar« and »Sonic Artefacts«, respectively. Smith has taken the opportunity to call out his fellow sound scholars to make their ears wet and listen to the sounds of urban water more carefully. Pisters brings her expertise on film theory to the floor to tackle issues such as the use of film music as expressionist sound, gendered representations of voice and the territorializing effects of sound. In response to Birdsall's chapter, Karanathasopoulou and Crisell suggest to move beyond studying radio for how it represents the city through sound and equally focus on how it has contributed to an urban aesthetics more widely.

The three main chapters and their expansions constitute the first three parts of our book. The last, fourth part harbours three contributions that each reflect on our book's key theme in a rather fundamental way. Sound and Media Studies scholar Jonathan Sterne critically takes up the longstanding debate on the virtues and vices of the notion of »soundscape« itself. Interestingly, he sheds new light on the work by the author to which R. Murray Schafer himself credits the notion of soundscape, and how this work informed Schafer's take on it. But even more importantly, he contextualizes the notion by linking it to the rise of hifi stereo sound in the 1960s. This justifies, a posteriori, our use of the concept of soundscape in a book about the staging of sound in media productions: speaking about soundscapes is particularly topical and useful when analyzing *mediations* of our sonic environments.

In the subsequent contribution, anthropologist of auditory culture Holger Schulze unravels and reflects on the experience of listening to museum audio guides. Rather than discussing the content of specific audio guides, he draws on phenomenology, anthropol-

ogy and the notion of *dispositif* to clarify what happens with our body when we listen to audio guides. He shows, in fact, how complex it is to wander through a museum and orient ourselves in time and space when also listening to recorded sound.
Our book ends with silence. In the last chapter, Ross Brown discusses the two minutes of silence on the Eleventh of the Eleventh of the Eleventh. It is in these two minutes, in London as well as in similar rituals in Amsterdam and other Western cities, that we experience the city anew by listening to its silence. Radio and televised recordings and broadcasts of these silences in fact stage city noises once again, albeit in subdued and perhaps even more telling ways.

5. Acknowledgements and Concluding Remark

The research project from which this book originated has been funded by the *Netherlands Organization for Scientific Research (NWO)* in the context of its research programme on Cultural Dynamics. We would like to thank NWO for its interest in our idea to study mediated sound as an alternative and intriguing entrance into the study of cultural change. We are also highly grateful to the *Faculty of Arts & Social Sciences of Maastricht University* for enabling us to organize the workshop Soundscapes of the Urban Past, where we discussed the chapters, expansions and reflections for this book. In addition, we would like to express our gratitude to photographer Moniek Wegdam. On February 13, 2012, she took the photo that is now on our book's cover. It shows the Dam, Amsterdam's most famous square, from exactly the same point of view as George Hendrik Breitner painted it in the years between 1895-1898. Feel free to compare the photograph from 2012 with the painting from the late 1890s. It will give you a first, highly mediated impression of the differences between the Amsterdam soundscapes of today and those of the late nineteenth century. What do you think: was it more silent back then? Or should we listen to the sounds as mediated by our cultural heritage rather differently? This concern is the topic of our next chapter.

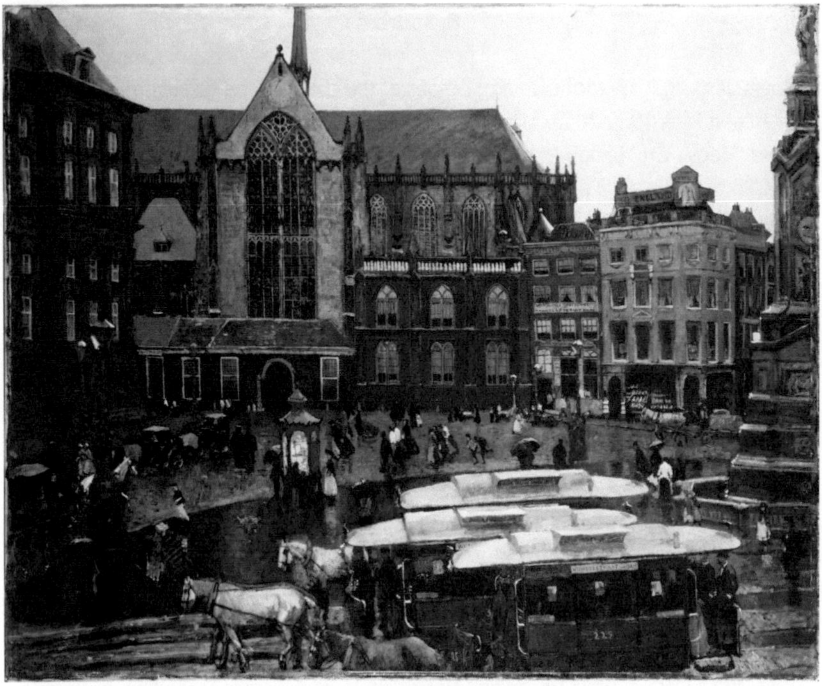

Figure 1: George Hendrik Breitner, *Gezicht op de Dam* (Amsterdam, 1895-1898)

6. Sources

Aalbers, Jasper (2013, forthcoming): Echoes of the City. Staging Urban Soundscapes in Fiction Film, 1930-2000, Maastricht.

Adams, Mags / Guy, Simon (eds.) (2007): Special Issue The Senses and the City, in: The Senses & Society 2.2.

Bijsterveld, Karin (2001): The Diabolical Symphony of the Mechanical Age. Technology and Symbolism of Sound in European and North American Noise Abatement Campaigns, 1900-40, in: Social Studies of Science 31.1, 37-70.

Bijsterveld, Karin (2003): ›The City of Din‹. Decibels, Noise and Neighbors in the Netherlands, 1910-1980, in: Osiris 18, 173-193.

Bijsterveld, Karin (2008): Mechanical Sound. Technology, Culture and Public Problems of Noise in the Twentieth Century, Cambridge MA.

Birdsall, Carolyn (2012): Nazi Soundscapes. Sound, Technology and Urban Space in Germany, 1933-1945, Amsterdam.

Brady, Erika (1999): A Spiral Way. How the Phonograph Changed Ethnography, Jackson MS.

Bull, Michael / Back, Les (eds.) (2003): The Auditory Culture Reader, Oxford.

Cockayne, Emily (2007): Hubbub – Filth, Noise & Stench in England, 1600-1770, London.

Corbin, Alain (1995): Time, Desire and Horror. Towards a History of the Senses, Cambridge.

Corbin, Alain (1999): Village Bells. Sound and Meaning in the Nineteenth-Century French Countryside, London (Originally published in 1994 as Les cloches de la terre. Paysage sonore et culture sensible dans les campagnes au XIXe siècle, Paris).

Cory, Mark E. (1994): Soundplay. The polyphonous tradition of German radio art, in: Kahn, Douglas / Whitehead, Gregory (eds.), Wireless Imagination. Sound, Radio, and the Avant-Garde, Cambridge and London, 333-71.

Cowan, Alexander / Steward, Jill (eds.) (2007): The City and the Senses. Urban Culture Since 1500, Aldershot.

Cowan, Michael (2006): Imagining Modernity Through the Ear, in: Arcadia 41.1, 124-46.

Diaconu, Mădălina / Heuberger, Eva / Mateus-Berr, Ruth / Vosicky, Lukas Marcel (eds.) (2011): Senses and the City. An Interdisciplinary Approach to Urban Sensescapes, Vienna.

Döhl, Richard (1985): Das Neue Hörspiel. Geschichte und Typologie des Hörspiels, Darmstadt.

Goodale, Greg (2010): Sonic Persuasion, Reading Sound in the Recording Age, Champaign IL.

Howes, David (2011): Vienna. Sensory Capital, in: Mădălina Diaconu / Eva Heuberger / Ruth Mateus-Berr / Lukas Marcel Vosicky (eds.), Senses and the City. An Interdisciplinary Approach to Urban Sensescapes, Vienna, 63-75.

Jacobs, Annelies (2013 forthcoming): Het geluid van gisteren. Waarom Amsterdam vroeger ook niet stil was, Maastricht.

Krug, Hans-Jürgen (2003): Kleine Geschichte des Hörspiels, Konstanz.

Lastra, James (2000): Sound Technology and the American Cinema. Perception, Representation, Modernity, New York.

Meszaros, Beth (2005): Infernal Sound Cues. Aural Geographies and the Politics of Noise, in: Modern Drama 48.1, 118-131.

Morat, Daniel / Bartlitz, Christine / Kirsch, Jan-Holger (eds.) (2012): Politik und Kultur des Klangs im 20. Jahrhundert, Zeithistorische Forschungen 8.2 (special issue).

Müller, Jürgen (2011): "The Sound of Silence". Von der Unhörbarkeit der Vergangenheit zur Geschichte des Hörens, in Historische Zeitschrift 292, 1-29.

Parr, J. (2010). Sensing Changes. Technologies, Environments, and the Everyday, 1953-2003. Vancouver / Toronto: UBC Press / Seattle: University of Washington Press.

Picker, John M. (2003): Victorian Soundscapes, Oxford.

Pinch, Trevor / Bijsterveld, Karin (eds.) (2012): The Oxford Handbook of Sound Studies, Oxford.

Pisters, Patricia (2007): Photographs of People Talking? Visies op geluid, in: Pisters, Patricia, Lessen van Hitchcock. Een inleiding in mediatheorie, Amsterdam, 165-89.

Rajewsky, Irina O. (2005): Intermediality, Intertextuality, and Remediation. A Literary Perspective, in: Intermédialités 6. Fall, 3-64.

Rath, Richard Cullen (2003): How Early America Sounded, Ithaca / London.

Schafer, Raymond Murray (1994): The Soundscape. Our Sonic Environment and the Tuning of the World, Rochester Vermont (originally published in 1977 as The Tuning of the World, New York).

Schwartz, Hillel (2011). Making Noise. From Babel to the Big Band and Beyond, New York.

Schweighauser, Philipp (2006): The Noises of American Literature, 1890-1985. Toward a History of Literary Acoustics, Gainesville FL.

Smith, Mark M. (2001): Listening to Nineteenth-Century America, Chapel Hill.

Smith, Mark M. (ed.) (2004): Hearing History. A Reader, Athens.

Sterne, Jonathan (2003): The Audible Past. Cultural Origins of Sound Reproduction, Durham.

Steward, Jill / Cowan, Alexander (eds.) (2007): Introduction. In Alexander Cowan / Jill Steward / (Eds.), The City and the Senses. Urban Culture Since 1500, Aldershot, 1-22.

Thomsen, Christian W. / Schneider, Irmela (1985): Grundzüge der Geschichte des europäischen Hörspiels, Darmstadt.
Thompson, Emily (2002): The Soundscape of Modernity. Architectural Acoustics 1900-1933, Cambridge MA.
Thuillier, Guy (1977): Pour une Histoire du Quotidien au XIXe siècle en Nivernais, Paris.
Zuyderfelt, Rutger (2009): Take a Closer Listen, Rotterdam.

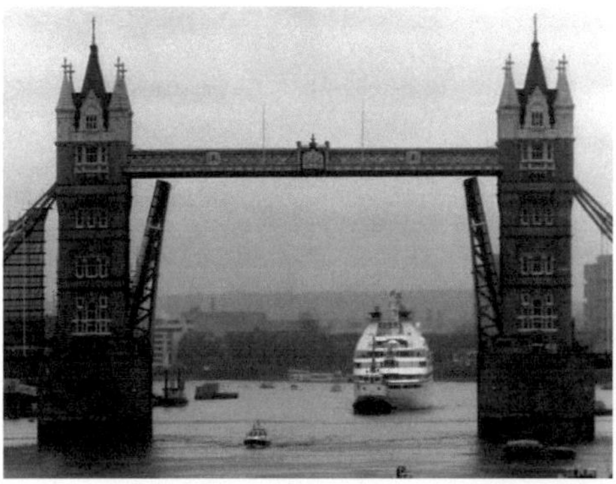

Shifting Sounds

Textualization and Dramatization of Urban Soundscapes

Karin Bijsterveld, Annelies Jacobs, Jasper Aalbers and Andreas Fickers

1. Sound Ways in the Museum

In 2007, the *Aboa Vetus & Ars Nova Museum* in Turku, Finland did something remarkable. It organized an exhibition on the history of the medieval town of Arboa Vetus which hardly made use of images, while also very few artefacts were put on display. The exhibition, entitled »Sound Ways«, was largely built around sound and text. As such it re-enacted the acoustic environment of Arboa Vetus by playing recordings, or »acoustic images« as the curators called them, in the rather dark museum basement harbouring the rudimentary remains of the town's excavations. These recordings presented six medieval characters: a blacksmith, a fisherman's wife, a cathedral schoolmaster, a maid of a merchant's house and a mother superior. Rather than being played over headphones, the recordings automatically started to sound as soon as visitors entered a particular area of the exhibition. To prevent a cacophony of sounds these spaces were separated from each other through sound isolating felt.[1]

In looking back on this exhibition, the curators explained that they aimed to find a solution for an issue we would like to label as the »museum authenticity problem«.[2] For some, this problem refers to the seeming contradiction between, on the one hand, the academic claim that historical authenticity is constructed by the observer rather than »inherent or intrinsic to the person or practice or object« presented and, on the other hand, the firm belief of the wider public to simply experience authenticity through these presentations, for instance by listening to early music played with historical

1 Aboa Vetus & Ars Nova Museum 2007.

2 Lehto-Vahtera 2010, Muhonen 2010.

accuracy.[3] What the Sound Ways' curators had in mind, however, was the problem that museums aim to bring their visitors into *direct* contact with the past by showing them »authentic objects«, while for all sorts of reasons these objects have increasingly become *untouchable*. They have not only been glass-cased due to issues of security and preservation, but in many instances they are also presented in a highly aestheticizing manner: as objects in splendid isolation that have to express their »aura« as well as the genius of their makers. Yet this practice also distances the object from the visitor: it is behind glass, in the spotlight, on stage and not to be touched.

One potential way around preservation and security issues is to visually mediate the object by using photos, copies or internet presentations – these can be touched without harming the original in any way. Yet such strategies exactly remove the objects' aura, even more so if the original object involves sound, such as that of bells or gramophones. »Auratizing« or »re-auratizing« the experience of past objects and surroundings, here in its double meaning of making things *aura*ble and *audi*ble, could be a more sustainable solution to this problem.[4] Given that we are used to rely in part on our ears when it comes to deriving information about spaces and objects, sounds – recorded or natural – have a quite powerful »material« and »tactile« dimension. They may create the feeling that the sounding object is close to you even if it is not there, or that you are really immersed in a particular space even though the sounds it generates were recorded.[5] And if a sounding object *is* in fact present in the museum, it does not necessarily need to be touched by the visitor to make it sound, thus rendering it audibly tangible.

This was precisely what Sound Ways wanted to accomplish. For example, this exhibition gave visitors a feel of parchment by playing the *sound* of parchment when they turned a page of a fake medieval tome. In addition, the exhibition sought to convey the meaning of sound by explaining how in the medieval era people attributed magical forces to particular sounds. Finally, the makers of the exhibition helped the contemporary visitor to empathize with his medieval forerunner by presenting auditory characters, people coming to life through a reconstruction of their sounds, and making use of audiographs – a consistent acoustic profiling through voice and accent (see Chapter One), even though the Sound Ways' curators did not actually use these terms when writing about their exhibition.

3 Titon 2008.

4 For an explanation of Walter Benjamin's notions of »aura« and »reauratisiering«, see Leschke 2003, 172. Using slightly different wordings and arguments than we use here, Andreas Fickers identified the museum authenticity problem and some of its solutions in Fickers 2012.

5 About the tactile dimension of sound and its effects in the museum context: Marsh 1998, 55.

Sound Ways was highly original in its approach. The use of sound in museums as such was not, however. Over the past three decades, the sonic realm has become an increasingly important aspect of exhibitions, in the form of »narrative, music and ambience«.[6] The use of audio guides, podcasts and sound walks, for instance, has become highly common (see Holger Schulze's contribution to this volume).[7] In addition, historical museums use recorded sound and music for creating a particular atmosphere, for recreating historical soundscapes (as in Sound Ways, or in the aural battle scene of World War I in the *Flanders Fields Museum* at Ieper, Belgium[8]), for collections of vanished sounds (on computer displays)[9] and for interactive, virtual soundscapes. The *Ruhr Museum* in Essen, for instance, has employed interactive sound by creating an area in which visitors can start recordings of particular mining industry-related sounds by positioning themselves on a particular spot under a sound shower.[10] Visitors stepping from one spot to another or coordinating how each of them is positioned thus build up their own soundscape composition of sounds associated with mines. In the exhibition *Sound Souvenirs* in Museum *Het Domein* in Sittard, the Netherlands, children were invited to do a sound memory game in which they had to find matching sounds by opening the drawers of a cabinet.[11] And *Das Haus der Musik* in Vienna offers visitors the option to create their own sound art compact disk by sampling and combining sounds recorded in urban and other settings.[12]

The use of sound in exhibitions comes with problems of its own, however. We already hinted at one particularly pressing problem: noise pollution in the museum. Listening with headphones may seem an appropriate answer, if some visitors would not complain about feeling »isolated [...] from the museum environment«.[13] Another troubling issue pertains to the attention span of visitors and the time and serenity it takes to wait and listen; for this reason, the visitors of Sound Ways were invited to *lay down* for a couple of minutes. A more fundamental problem, identified by Mark M. Smith, is that the presentation of the sounds of the past in museum settings may be too straightforward. He mentions the example of an American cotton mill museum that stages early nineteenth-

6 Stocker 1994, 178.

7 About the use of podcasts in museums: Yasko 2006.

8 In Flanders Fields Museum.

9 Lane / Parry 2003.

10 Ruhrmuseum.

11 Museum Het Domain, 2007-08.

12 Haus der Musik.

13 Angliss 2005, 29.

century industrial sound simply as »loud«. Such an approach will hardly convey »the full texture« of sounds or »the context in which people, especially those on the shop floor listened to « these sounds.[14] Contextualization is even more important when museum visitors listen to sound without any visual reference. As they immediately start guessing about what they listen to, they tend to hear what they are familiar with. Presenting sounds without telling what they meant to past listeners thus comes with the risk of historical inaccuracy, while telling what past listeners heard without presenting the sounds themselves may be disappointing in terms of the museum experience.

And yet many museums appear to adopt either the first option (sounds without much explanation) or the second one (explanation without much sound), as Holger Schulze has noted in his contribution to this volume. It is possible and interesting to do both, however, even though »connecting with the life of the senses in different historical periods« is »no easy task in view of the ephemeral nature of the sensate«, as David Howes has rightly claimed.[15] It is exactly for this reason that the staging of sound in text, radio or film, which in this chapter we illustrate through the example of urban sound, is potentially highly informative and enriching in museum contexts.

2. Capturing and Staging the Experience of Sound

Let us return for a moment to the stories on favourite sounds gathered by Rutger Zuyderfelt and those on unwanted sounds collected by the *Groningen anti-noise committee* – the stories presented in the introductory chapter to this volume. We identified the first set of stories as narratives that often staged sound by employing the auditory topos of the comforting sound, and the second as letters commonly using the auditory topos of the intrusive sound – including its stylistic characteristics such as rhythmic enumeration – to articulate particular experiences of urban sound.

Comforting or intrusive, what the stories have in common is that they were solicited by highlighting »novelty« in the experience of the urban sonic environment. Zuyderfelt invited people to listen to their sonic environments *as if their Ipods were broken*, and as if outdoor noises could sound like a recorded composition. The members of the anti-noise committee used their press reports to provoke responses from their fellow citizens, and they analyzed these responses in detail to further substantiate their conviction that city

14 Smith 2012, 39, 40.

15 Howes 2011, 63.

life had recently transformed into a deafening experience. In both situations, the inviting actors aimed to heighten the public's auditory consciousness by making the sonic environment sound »strange« or »new«. Our thesis in this chapter is that this is exactly the narrative strategy often used in staging sound in historical text, radio and film – our cultural heritage of mediated sound.

Writers, radio-makers and filmmakers often deploy shifting plot situations as the backdrop against which to present and articulate »urban soundscapes«, here understood as both the sonic environments of city residents *and* the ways in which they make sense of the sounds they are surrounded with.[16] For example, the protagonist's arrival in a city unknown to him represents such conventional moment for the sonic display of the city in movies, radio plays and texts. At such »moment«, the sounds and the novelty of the setting intriguingly highlight each other: the sounds signify what is new and specific about the city to the protagonist, while simultaneously the narrative novelty of the situation makes the mundane sounds of everyday city life »audible« to protagonist and audience alike. This is not meant to suggest that there is no sound in the remainder of the play, movie or story; rather, this strategy presents makers with a chance to zoom in on the sonic characteristics of some urban environment at some point in time, or on how the protagonist perceives them.

This chapter, then, is not about »shifting sounds« in the sense of »sounds changing over time«, even though the reader will encounter many a modulated or vanished sound when reading our essay. Instead, it focuses on the narrative repertoires used in text, film and radio to articulate particular impressions of urban soundscapes. As we will show, intense and contrasting shifts in time and space help to capture and stage the specificities of soundscapes and their perception by the protagonists. In most cases it is quite challenging to capture fleeting sensory experiences tied to the past or present in words, images or sounds convincingly. A recent study on people's sensory experiences of family resemblances revealed that photo elicitation may stimulate respondents to talk about such experiences in a rich vocabulary, but the researchers also ran into many »epiphanal« stories of people struggling with the short-lived nature of their »sensory experiences. Several respondents indicated that they might have seen, heard or felt something at one point in their life very briefly, but that soon after they lost access to it again.[17] Similarly, artists struggle to capture the ephemeral nature of our sensory life. Even the artist-composer Luigi Russolo, a Futurist who aimed to transform and enrich classical music with the microtones and timbres akin to the urban sounds of his age –

16 We follow the definition of »soundscape« by Emily Thompson, see Thompson 2002, 1.

17 Mason / Davies 2009, 596-97.

the early twentieth century – was heavily criticized for not adequately capturing the essence of urban sound. As a Dutch reviewer wrote about the sound of Russolo's new musical instruments, the *intonarumori*:

> Yet it remained a gray, opaque, chromatically up-and-down-whining rush. Nothing of the machine factory's majestic, dark-toned chanting. Nothing of the rhythmic *pointillé* of ship hammering, or a simple typewriter. Those searching for things to compare the auditory impressions with did not get any further than clearing one's throat, a dentist's drill, or, even worse, a potpourri of lavatory sounds [...] The Futurists really seem to be on the wrong track with their naturalistic imitation of modern city noise, and are even surpassed by the popular jazz band that, with its overpowering hubbub-avalanche and panting rhythm, gives a much more convincing transposition of our modern, cinematic sensation-life![18]

This chapter will explore how others have tried to »catch« the fleeting auditory experiences of the city. It is about recurrent, conventional ways of staging urban sounds – ways in which stark narrative transitions in time and space function as key elements, even though we also employ the analytical tools presented in our introductory chapter: auditory topoi, keynote sounds, soundmarks, sonic icons, audiographs and acoustic profiling. Of course, our analysis of such narrative tools for staging sound is hardly exhaustive and cannot be anything but preliminary and exploratory at this point. As such our effort is confined to the sources collected for our earlier studies on staging sound in historical texts, radio plays and films (see Chapter One, footnote 21). Moreover, in this chapter we focus more on examples from text and film than from radio, because radio is the key topic of the third part of this volume. That said, we will show how writers, radio play makers and filmmakers have employed narrative shifts in time and location to present the sounds of the city in ways they considered relevant and convincing. We will discuss three of these narrative transitions in particular: arriving in the city, following the rhythm of the urban day and juxtaposing soundscapes in terms of either place (different neighbourhoods within a city) or time (past and contemporary soundscapes). After a brief, critical consideration of an issue easily overlooked when focusing on these narratives – the issue of adaptation to sound – we will return, in our conclusion, to our initial concern with how these narrative strategies for capturing the experience of sound can be informative to curators and others involved in making historical exhibitions which employ sound.

18 Anonymous 1921, 6.

3. Arriving in the City

Arrival scenes are common in all sorts of media. One of the most famous *urban arrivals* in early film history is the opening scene of Phil Jutzi's *Berlin Alexanderplatz* (1931), a film after Alfred Döblin's novel of the same title (1929). Protagonist Franz Biberkopf has just been released from prison, but his freedom does not only bring him joy: the sounds of the city beyond the walls of the prison frighten him. When he boards a tram headed for Alexanderplatz, we hear non-diegetic music that gradually merges with the sounds of the street: the noises of the tram engine, of road construction workers hitting stones without a clear rhythm, of tram bells incessantly ringing and cars suddenly sounding their horns. We see rapidly passing city scenes from the perspective of tram passenger Biberkopf. Simultaneously, the traffic sounds turn louder and louder, eventually over-crowding the music – resulting in what Jasper Aalbers has coined the »noisification of music«.[19] The scene reaches its climax when the protagonist jumps off the tram: the sounds and sights of the city are simply too much for him.

The traffic sounds invoke Biberkopf's nervousness, prompting him to leave the tram and to search for tranquillity in a courtyard. The shift in location from a presumably boring and silent prison to the noisy inner city articulates the trope of the intrusive sound, of noise the protagonist cannot immediately cope with, over-stimulated as his senses are. The next scene, which marks a shift from the noisy tram into the quiet courtyard does exactly the opposite: it brings in the trope of the comforting sound. Intriguingly, the traffic sounds express both the character of the city – a busy, frantic and threatening agent – and that of Franz Biberkopf's state of mind. As the German film historian Guntram Vogt put it:

> The transition from hectic traffic to the seclusion of an inner courtyard [...] and the atmosphere of the metropolis allow for a striking expression of the character's psyche.[20]

Importantly, the above-described film scenes proved extremely hard to record. The microphones used at the time were so sensitive that it was impossible, for instance, to film the scene *and* to record the dialogue on location. For this reason, Alexanderplatz and its vicinity had to be rebuilt at the Babelsberg film studios.[21] The background sounds for

19 Aalbers 2012, forthcoming.

20 Vogt 2001, 29, Hake 2004.

21 Rehhahn 1996, 222, Vogt 2001, 252, Braunger 2008, 28.

the scenes situated at or near Alexanderpatz came from original recordings of 1930s traffic.[22] Apparently, the prerecorded, »standard« sounds from studio catalogues were deemed of insufficient quality to fulfil the important functions of urban sounds in *Berlin Alexanderplatz*.

The use of urban sounds in the *Berlin Alexanderplatz* novel, film and radio play, including a comparison between these versions, warrant a study in its own right, which we undertake in our chapter on »Familiar Sounds«, this volume.[23] Here, however, it is relevant to show how »changing places« equally opens up the protagonist's ears in many of our other examples from film, allowing the audience to listen in to a particular take on urban sounds. As the protagonist enters a city, the film audience arrives there together with him, joining his first experience of the urban environment. The opening sequence of a lost version of *The 39 Steps* (1935) by Alfred Hitchcock, for example, is, as one critic argued

> a brilliant evocation of what a loud and busy city can be to a friendless outsider. We see a quick tumble of images from the city [...] with a barrage of city noises.[24]

In such openings, the use of visual and sonic icons helps the audience to locate the scene almost instantly. In Hitchcock's thriller *Foreign Correspondent* (1940), an American journalist travels to London to cover a peace conference. Just prior to his arrival by train, visualized by the smoke and steam of a train entering a railway station, we see an image of the Clock Tower of Westminster Palace. We now know the scene is London. Yet we do not hear the Tower's Big Ben. Only two minutes later, when we watch the journalist in his hotel room, we hear its bell. While it sounds, the journalist does a little dance-improvisation involving an umbrella and a bowler hat: clearly, this is England. And just in case it has slipped our mind, the sound of the Big Ben reminds us that we have arrived at a particular place in England: London.

The four-toned chime of the Big Ben acts as an auditory landmark, a soundmark of London. By 1940, the shot of the Clock Tower had already transformed into a *visual* icon of London-in-film because the production company London Film used it as its logo from the early 1930s onward.[25] Similarly, the tower's chime developed into a *sonic* icon. In radio plays set in London for instance, such as *If you're glad, i'll be frank* (Tom Stoppard

22 Braunger 2008, 40.

23 See also Jelavich 2006.

24 Maurice Yacowar, quoted in Barr 1999, 20.

25 Brunsdon 2007, 12, 28.

1966) and *The dog it was that died* (BBC 1982), the sound of the Big Ben was used as both a spatial reference and a reminder of the inevitable march of time.
In *Trainspotting* (Danny Boyle, 1996), to return to examples from film, the Scottish protagonist Renton, a former drug addict, arrives in London where he aims to start a new life. The city is introduced through shots of London as tourists know it, including, again, the Clock Tower, but we do not hear any urban sounds. Instead, the images are accompanied by the exciting pop music of Ice MC's *Think About The Way.* Although the music is non-diegetic, it succeeds in having the audience share Renton's experience of arriving in London, a city of big business and new opportunities – the trope of sensational sound. The film *London* (Patrick Keiller, 1994) similarly opens with an arrival in London, in this case of a man, a photographer on a cruise ship, returning home. The opening scene shows the ship passing the Tower Bridge, and we hear it sounding its horn. Such horns are the sonic icons of London East End, where the docks used to be. This gives the sound a nostalgic and melancholic quality – in line with the overall mood of the film – as it refers to activities that no longer exist.

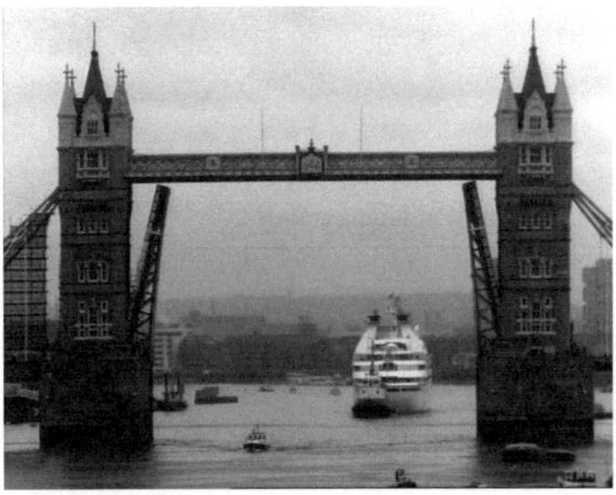

Figure 2: Still of the film *London* (1994) by Patrick Keiller.

But what if the film's characters or the audience are not familiar with the geography and sounds of London? In *Out of the Clouds* (Basil Dearden, 1955), a romantic comedy, two travellers – one from Germany and the other from the United States – are stuck in London because the airport had been shut down due to bad weather. They spend the night in a London pub, and while standing on a balcony they hear the sounds of ship horns and church bells. The pub's female owner explains to them what they hear:

> Just you stand and listen to it though. [*Ship horns are audible at a distance.*] I often come out here and listen. Like a concert it is sometimes, the sounds of the ol' River. [*A church bell rings.*] That's St. Mary's Rotherhithe, she's always the first.

> You'll hear all the others in a minute. Southwick and Bow and St. Paul's, and the City churches. [*Several church bells ring, one of them being the Big Ben, and a ship sounds its horn.*] There's a big'un coming up. That'll be the Dutchman I expect, always in on Fridays. [*A big rumble startles the tourists.*] That's just the barges banging together at their moorings; the wash of a big ship always sets them off.

There is a remarkable difference in how the various characters are shown to experience this scene. The English hostess enjoys the sounds of the city and the river as an agreeable concert: they are familiar sounds to her, which structure the flow of time. To the German and American tourists, however, the sounds are new, even startling. Only after the hostess has given these sounds a meaning and a location, these iconic London sounds create such a romantic atmosphere that as soon as she leaves to prepare for dinner, the tourists have their first passionate kiss.

Another interesting diversity in the meaning of the sounds of London to the protagonists is found in the opening scene of the radio production *On the town: London* (BBC, 1978). This radio portrait of London starts with twelve strokes of the Big Ben, and every interval between a chime is filled either by a recorded statement of what London means to an interviewed person or by a selection of sounds such as traffic noises, the murmuring of voices emanating from an anonymous crowd and the discrete sounds of birds in a park. While the statements between the chimes offer a great variety of views about London as a place loved or hated, pleasant or frightening, amusing or strange, the recurring sound of the Big Ben works as an acoustic and narrative frame, tying together the diverging characterizations of the metropolis by the protagonists.

As Annelies Jacobs has shown, travel narratives written by foreign visitors of late nineteenth-century Amsterdam similarly managed to open up the ears of their audiences by articulating how different the Dutch capital sounded from other European cities.[26] Much of the evidence about the sensory environment of early modern cities, Jill Steward and Alexander Cowan have claimed, »has come from outsiders, sensitive to material and cultural differences and eager to make comparisons«.[27] This was still true for the nineteenth century, when visitors listened to Amsterdam with fresh ears, much in the same way as a former Amsterdam resident who had moved to Groningen listened to his new town only to find its noise »striking«, unlike »perhaps the people who lived in Groningen all of their life and have grown more used to it«.[28] Travellers, it seems,

26 Jacobs 2013, forthcoming.

27 Steward / Cowan 2007, 3.

28 Archives Anti-Lawaai comité Groningen, 66.

listen carefully and may observe sharply by comparing, consciously or not, the sounds of newly visited cities with the familiar sounds of their hometowns.
What did these travel writers have to tell about the urban sounds of Amsterdam in the late nineteenth century? Although visitors such as the Italian Edmondo de Amicis, the Portuguese Ramalho Ortigão, the British Charles W. Wood, the French Henry Harvard and the Belgian Charles de Coster gave lengthy descriptions of such divergent sounds as local voices, bells, carillons, horse hoofs, carriages, carts and handcarts on the city's cobblestones, they had much less to say about the sounds of industry – with the marked exception of the Amsterdam diamond industry. Ramalho Ortigão considered the »ear-deafening grinding« of cleaving diamonds »nerve racking« – »irritating every single pore of our skin«.[29]
Most of the descriptions by foreign visitors focused on street sounds, however, notably the sounds of transport and traffic, including pedestrians. About a typical Amsterdam street, Charles Wood said that it

> was very narrow [...], had been paved with unequal stones, and reverberated intensely. A carriage that passed through it rattled with a violence as if it was a long artillery train.[30]

He had even worse noises to report about. At night, Wood felt not only disturbed by the many Amsterdam bells, but also by the nightly uproar of »loudly drunk men and women of the meanest kind«. In fact, these drunks »had voices as only the most uncivilized classes can have, or as only the Dutch posses«.[31] While he wondered about the sounds of Amsterdam at large, his fellow travel writer Charles de Coster tried to record in written words the characteristic sounds of one particular Amsterdam neighbourhood, the Jewish quarter: »It is hard to imagine the din and noise of the *Jodenbreestraat* if one has never actually heard it«, he claimed.

> All these small merchants cry, gesticulate, smile to attract customers, mention prices, sell, announce what they sell or buy and recommend its qualities. And what voices! These are nose and throat sounds in all keys. The small ones call in major scales and the big ones in minor. These qualified sellers use every sales trick imaginable. They attract attention by using rattles, castanets or by drumming on

29 Ortigão / Jong 1948 [1885], 150.

30 Wood / Baarslag 2002 [1878], 55.

31 Wood / Baarslag 2002 [1878], 54-55.

> old pans. And there is another rattle, a very big one, the city rattle. A Jewish boy turns it with great sincerity. He even plays a military march with it. In this manner he informs the housewives about the near arrival of the closed car that fetches household refuge at every door. A humpbacked Jew has harnessed himself before a cart with peat. On the peat sits a little boy who mimics a trumpet and invitingly beckons to passers-by. […] [A] boy with warped legs walks in front of it and tries to sell his ware by clattering a box with nails. An organ plays *Mère Angot* and another plays the lamenting song of the *Leeuw van Nederland*, loudly sang along by a few men and a woman in rags.[32]

Before explaining the specific wording used to stage to street sounds of the Jewish quarter and other areas of Amsterdam, let us first point to two other types of sound outsiders considered characteristic of Amsterdam. What they considered downright remarkable were the sounds of cleaning. On Saturday, the maids of Amsterdam turned houses inside-out:

> In residential housing, storehouses, shops, offices, everywhere everything is overthrown, dusted down, beaten, brushed clean, with a diligence, a fury, a fanaticism that is close to delirium.[33]

Also surprising to them was the right of Amsterdam children to bang their drums as loud as they wanted near the Stock Exchange during the week of the fun fair. As tradition had it, an orphan boy had discovered a »gunpowder treachery« under the old Stock Exchange, and by way of reward had asked and been given the eternal right for young kids to bang drums during the fair. In practice, this right was abolished when the *Beurs of Berlage* (the new Stock Exchange building) came into use in 1903, yet until that time, as Amsterdam commentators wrote, the event sounded like »disharmonic and ear-deafening violence«.[34] It seemed to shock Wood even deeper:

> When I left the New Church that morning and crossed the Dam (the most important square in town) I was surprised to see the Stock Exchange full of boys, and to hear emanating from it the sound of joy and laughter, cries, hurrays, drums, trumpets, flutes and sharp pocket flutes, in a manner I had never observed before and would not have been able to imagine, and of a type I hope never to hear again. In

32 Coster / Janson 1998 [± 1879], 36-37.

33 Ortigão / Jong 1948 [1885], 36.

34 Gouw 1974 [1875], 63.

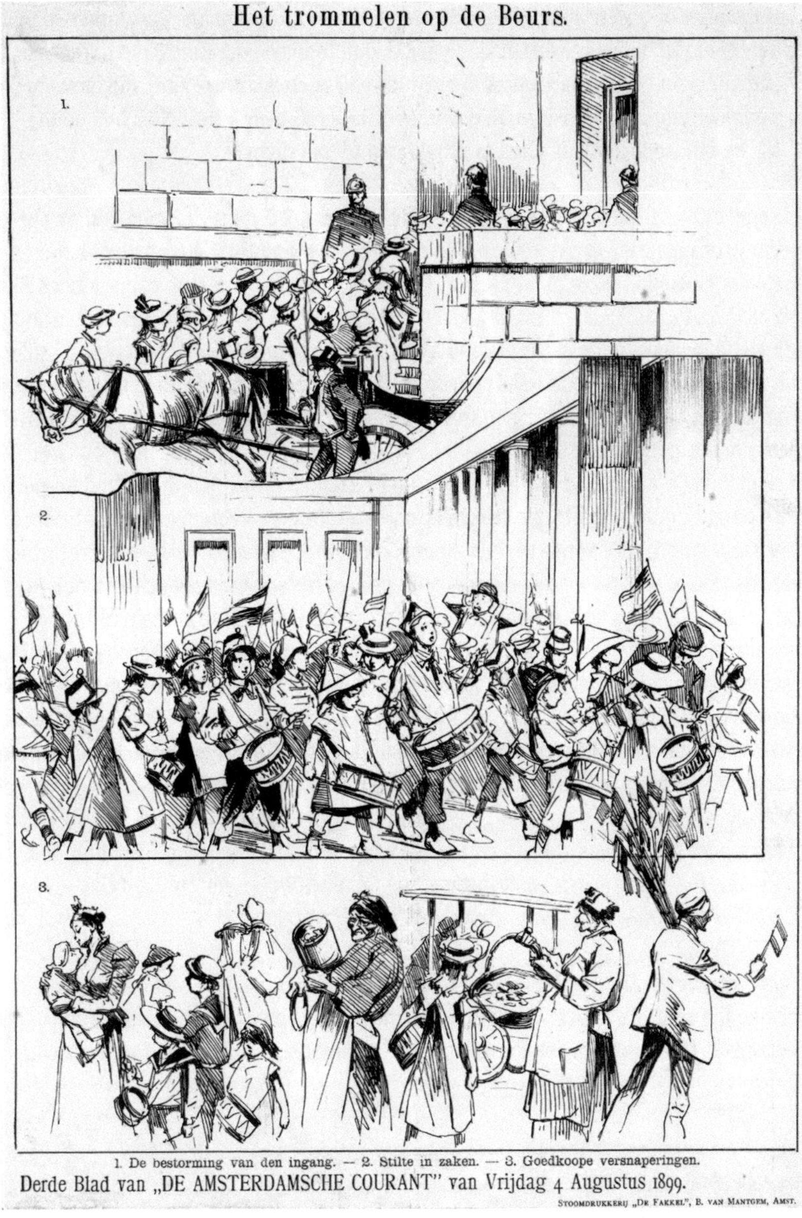

Figure 3: Banging the Drums at the Amsterdam Stock Exchange.

> a moment of absent-mindedness, I entered the Stock Exchange. Each and every corner of the building was chokingly filled with toddlers. Hundreds and hundreds screamed and yelled as loud as their Dutch voices could, drumming and blowing on ugly trumpets, ran like small devils, and clattered their little rattles behind the back of those incautious persons who dared to join them.[35]

As some of these quotes illustrate, observers tended to describe the sounds they heard in terms of musical language: not only did they give ample attention to the musical instruments used and songs produced on the streets, they also employed notions like »disharmonic«, »sounds in all keys«, »major«, »minor«. As Jacobs has convincingly argued, following Jonathan Sterne and Matthias Riege, this phenomenon was anything but a coincidence.[36] It expressed a way of thinking about sound that predominated until the late nineteenth century. In this line of thought, which started from Pythagoras' theory of harmony, music was thought to be fundamentally different from other kinds of sound: while music had regular or consonant sound waves, non-music had irregular or dissonant ones. In the 1860s, however, physicist Hermann von Helmholtz introduced new explanations for the rise of consonant and dissonant sounds, the by-effect of which was that music was no longer seen as principally *different* from speech and noise but considered to belong *with* speech and noise to one and the same domain of sound.[37] This also shifted scholars' priorities in *what* they aimed to examine. Prior to Helmholtz' insights the production of music – and thus the functioning of sources of music like the human voice, human mouth and musical instruments – were centre-stage. Afterwards, however, the effect and perception of music, and of sound at large, became much more important. In Jonathan Sterne's words:

> No longer themselves general categories of sound fit for theory construction, the mouth, the voice, music, and musical instruments would become specific contenders for audition in a whole world of sonic phenomena.[38]

It was the ear as such, and its ability to transform vibrations into audible signals, that became the central subject of research and the starting-point for new experiments and new sound technologies like the phonograph. Through the discussion of such sound

35 Wood / Baarslag 2002 [1878], 69.

36 Jacobs 2013, forthcoming.

37 Rieger 2003, 183-89.

38 Sterne 2003, 33.

technologies in the general press and the public's experience with them, novel ways of thinking about sound would reach the general audience.[39] This involved a process that took many years, however. A quote from a Dutch newspaper article on the phonograph reveals the still deep-rooted distinction between music and non-music in the late 1870s. The phonograph, as the article explains,

> clearly returns each sound, of whatever character. It does not make a difference whether or not it is a musical note, the cough of an old man, the sweet twittering twinkle of a Patti, the loud bark of a dog, or the screams of a parrot; once it has once been written on the phonograph's tinfoil, it can be repeated as often as one wishes.[40]

It is within this context that we can understand why late nineteenth-century visitors arriving in Amsterdam had a special ear for the voices, bells and musical sources of sound audible on its streets. And within these categories, they clearly foregrounded the sounds they considered typical of Amsterdam (and different from what they were used to), such as those made by the drumming children, the hammering sounds of cleaning, and the reverberation of these sounds in the city's narrow streets. Moreover, their »period ear«[41] did not only imply that they were attuned to the (a-)musicality of sound; they were also quick to associate the less agreeable sounds with uncivilized behaviour or lower classes and cultures, as the quotes above illustrate.

So even though »arrival scenes« served as a common narrative strategy for writers, filmmakers and radio-makers to »capture« urban sounds, their »period ear« deeply influenced how they staged these sounds (and their take on the city) in more detail. Intriguingly, the rising interest in the issue of noise in the early twentieth century and the »noisification of music« by Futurists and avant-garde filmmakers had been preceded by an age which preferably reviewed sounds in terms of their musicality.

4. The Rhythm of Urban Life

So far, we have discussed shifts in locations, or change across *space*, as a narrative entrance into the staging of city sound or a protagonist's experience of particular urban

39 Rieger 2003, 193.

40 Anonymous 1878, 1.

41 Burstyn 1997.

sounds. The use of shifts in *time* is another important element of repertoires for staging sound. In this section we will show how evocation of the rhythm of everyday urban life has served as a common strategy for capturing the sonic experience of city life and staging particular characteristics of individual cities.

A most emblematic example of this staging strategy in fact involves a *silent* movie: Walter Ruttman's *Berlin: Symphonie einer Grossstadt* (1927). This film, which is entirely devoted to capturing the rhythm of life in the metropolis, opens with a train that enters the city in the morning, after which it shows how Berlin is awakening, how its traffic gets moving and how its citizens are going to work in huge factories. Images of crowds, traffic and industrial activity are accompanied by an energizing as well as slightly chaotic music score composed by Edmund Meisel. At the beginning of the fourth act, however, the music suddenly halts: time for lunch. Construction workers leave the shop floor and sit down for sausages and beer. As two horses are being fed, the music turns into a slow and relaxing tune, followed by funny scenes of an elephant in the zoo, picking up nuts with its trunk, and a moving puppet in a baker's shop window. The incessant hustle and bustle of city life has shifted into a slower pace – if only briefly.

Only a few years later, in 1930, Ruttmann produced *Weekend*, a purely sonic portrait of Berlin (see also the Chapter »Sounds Familiar«, this volume). The radio play had six movements and expressed the rhythm of a Berlin weekend starting on a Saturday afternoon at the shop floor, through a night out and a pastoral Sunday, and ending with the city returning to work on Monday. Ruttmann gathered his sounds by driving around Berlin in a van with a hidden microphone, stopping at locations such as train stations, factories and busy streets to record the uninhibited rhythm of city life.[42] For Ruttmann, *Weekend* was a study in sound-montage, in which »sound was an end itself«.[43] In 1998, the Bavarian public broadcasting service initiated a *Weekend remix* competition, which resulted in six different creative appropriations of the play by sound artists from all over Europe.[44] The French sound artist trio *To Rococo Rot* produced a remix in which they replaced Ruttmann's recordings by newly recorded sounds of a Berlin weekend in 1998. By keeping up the rhythmic montage principles of the original, they were able to produce another striking expression of Berlin's rhythm of life, yet this time of the late twentieth century.

Authors of written accounts of everyday life in late nineteenth-century Amsterdam also exploited the rhythm of urban life as stepping stone for staging sound. In capturing this

42 Shapins 2008.

43 Jerzy Toeplitz im Gespräch mit Walter Ruttmann (1933), in: Wiadomsci Literackie (28), 499, reprinted in Goergen 1989, 90.

44 Ruttmann, DJ Spooky et al. 1998.

rhythm, they evoked the density of urban life: the sheer number of people around and active at particular hotspots in the city – sensational sound once again. As the Frenchman Henry Havard claimed, the Kalverstraat, one of the main arteries of the Amsterdam business and shopping district, was »always full of life«, »no matter the time of the day«. Only »[b]etween 3 and 5 in the morning«, as he qualified his statement, »it may be quiet for a moment«.[45] To the Dutch chronicler Justus van Maurik, this was even true for Amsterdam at large. A city like Amsterdam, he claimed, »actually never gets entirely silent«. Even those living along the putatively quiet canals would always hear some distant sound, »reminding of the far-away murmur of the sea, weaker at some moment, stronger at another«.[46]

One frosty night, when the streets of Amsterdam were covered with snow, Van Maurik described what he heard while in bed. After »the old tower's clock« strikes twelve, he hears the distant sounds of other bells, a carriage and voices singing. This is followed by a carriage's wheels rattling on a bridge – it halts in front of the neighbours' home – and muffled voices, a door bell, the opening and closing of a coach door. After its leaving, the rattling of its wheels slowly dies down. Van Maurik falls asleep, but awakens before the clock struck six. Behind his house, in a narrow street,

> a knocker-up starts knocking, and he is making noise for two. This is where the craftsmen live, people who have to get up early.[47]

Next, he overhears farmers coming into town from all directions, first those with vegetables, who shortly afterwards are followed by again other types of cart that »rattle incessantly on bridges and streets«. Clearly, the »milk carts are back in the city«. By the time it is nine in the morning, »all city districts« are bustling with activity again, accompanied by occasional vendors' cries.[48]

Kalverstraat was particularly lively indeed. In 1875, an Amsterdam resident gave a detailed description of a typical day in this main shopping street:

> The first shift is that between 9 am and 1 pm. At that time, Kalverstraat is as calm and quiet as any other neighbourhood in town. First, one meets the children going to school […], devout Catholics going to church, gentlemen going to their office […].

45 Havard 1876, 461.

46 van Maurik ±1890, 217.

47 Knocker-ups could be hired to wake up those who had to start working early in the morning.

48 van Maurik ±1890, 1-9.

> Somewhat later, one finds the young ladies who do their shopping, strangers leaving their lodgings to see the capital's most remarkable sites, *commis-voyageurs* in a hurry, and organ-grinders turning against each other, playing the most ear-deafening music [...]. At noon, many a worker takes a break: the Jewish vendors with their carts with pears, apples, lemons or oranges; the fish vendors with their barrows; and the fishwives with their baskets full of living seafood. Things gradually come to a halt.
> The second shift lasts from 1 to 6 pm. First there is a flow of distinguished gentlemen and young men who hurry towards the Stock Exchange, which starts at half past 1. Next, you see women of doubtful reputation, whose attitude and dress reveal them to belong to the »demimonde« [...]. Now the *flâneurs* follow, with or without ladies; gentlemen who take their time to saunter up and down the street, take a halt in front of various shops, and, when they are on their own, take a rest from their fatigues in one or another coffee house, looking through the windows to watch, in turn, other *flâneurs* as well as *flâneuses*.
> Subsequently the flow of Stock Exchange visitors turns up again, most often in groups of two or three men talking excitedly with each other about the day's stock rates. Some of them enter a coffee house; others move on with this genuine Amsterdam footstep that tells you: »time is money«. Yet one will find fewer of them before long: many, whose offices are along the route, have secured a place in one or another omnibus. The streets are now growing ever more crowded, notably with pedestrians hurrying back to their homes.
> Between 5 and 7 pm *Kalverstraat* is nearly deserted. But soon those visiting the theatres, vaudeville or the people's palace start showing up: men and women, walking calmly or in haste, depending on whether they have reserved places or not. In their wake there are those who go for a walk, to catch a breath of fresh evening air or just for a stroll; the coffee houses get crowded.[49]

Actually, a year before this description was published, Kalverstraat was paved with asphalt, a novelty. Also unique at that time was that the city authorities only allowed traffic to move in one direction. The asphalt pavement was an experiment intended to reduce the noise of iron-mounted wheels on cobblestones, to raise the quality of the flaneur experience and to enhance the attraction of the shops. The new Kalverstraat sound had a downside, however. Because the police failed to enforce the obligatory use of bells for carts passing through, many people did not hear them approach anymore.[50]

49 Gouw 1974 [1875], 52-55.

50 Gouw 1974 [1875], 31.

Looking back in 1938, the Dutch classicist and literary scholar Aegidius Willem Timmerman claimed that the past had been noisier than the present:

> If we briefly ignore the sounding of car horns, the grinding of music disks, the whining of loudspeakers, towns of the past used to be much noisier than today's towns. No asphalt, no rubber tires, no ban on loud vending, [...] no fixed days for street organs or other musicians. Farmer carts without springs and lorries rattle-thundered on the rough round cobblestones, their sounds reverberating down the streets at two hundred meters' distance. From the early morning until the late evening, the cries of merchants selling [...] cantaloupes and oranges resonated. [...] Notably on Monday, throngs of drunks swarmed and swob through the city, arm in arm and screaming [...]. The place was teeming with street organs and street musicians. [...] You also heard the dustman rattling, hussars blaring, military music bands crossing the streets almost every day, rammers singing as the block fell, and the chinking of the chains.[51]

This commentary contradicts the common notion of city life growing ever noisier from the late nineteenth into the first half of the twentieth century, while it underlines how staging the nearly incessant sounds of city life by highlighting its rhythm was a convention in its own right. This passage also points to two other strategies for articulating a city's sonic realm: juxtaposition of present soundscapes with those of the past or future, and contrasting sounds among neighbourhoods within the same city, two topics we address in the section below.

5. Juxtaposing Soundscapes

»Way into the night«, the Italian traveler Edmondo De Amicis wrote in 1876, Kalverstraat

> swarms with men and women, gentlemen and ladies, in all kinds of mood, of all kinds of character; yet in just a few footsteps one leaves the illuminated street for the dark canals, between the motionless, idle ships, surrounded by a deep silence.[52]

51 Timmerman 1938, 257-58.

52 De Amicis et al. 1985 [1876], 196-97.

He effectively contrasted the sound of Amsterdam's business and entertainment district with that of the upper class neighbourhoods along the canals – without actually describing the first. In just one sentence, he captured a hallmark of the Amsterdam soundscape that may still strike visitors today: the luxurious tranquility along the canals with their aristocratic houses found in the heart of a buzzing city.

Jasper Aalbers identified several similar scenes of sonic contrast in films set in London and Berlin, although *urban arrival* scenes in movies more often triggered a characteristic representation of city sounds than did shifts from one neighbourhood to another, at least in his corpus. One interesting example is the role of children's sounds. In scenes set in the poorer districts of London, we more frequently hear children playing outside than in the fancy districts. In *Steptoe & Son* (Cliff Owen, 1973), for instance, children's sounds make up an important part of the keynote of scenes situated in or near the homes of two of the lead characters, who run a messy scrap iron business. And in *Nil by Mouth* (Gary Oldman, 1997), children are audibly playing amidst the sober, concrete blocks of apartment buildings. In contrast, children are conspicuously absent from the fancy streets of *Notting Hill* (Roger Michell, 1999), *Four Weddings and a Funeral* (Mike Newell, 1994) and *Bridget Jones' Diary* (Sharon Maguire, 2001). This contrast is played out in *Wonderland* (Michael Winterbottom, 1999), where we clearly hear children's activities near the cheap apartment of protagonist Nadia, while such sounds are absent in the more distinguished district where her parents live. The sounds of children playing outside has thus become a *keynote icon*, or conventionalized part of the background sound of a particular location, of poor, lower-class areas, where children presumably spend less time on organized, educational activities under the supervision of adults than in upscale neighbourhoods. It is also a form of acoustic profiling (see Chapter One), this time not only through the staging of the sounds of language and voice, but also through the sounds of a neighbourhood's activity pattern.

It is important to note, however, that in many movie scenes which portray transitions from one location to another within the same city, filmmakers use background sound or its absence to characterize the state of mind of the protagonist as well as the sonic environment. After the opening scene of *Wonderland*, in which Nadia sneaks away from a disappointing blind date, she wanders through the streets of Soho (presented in fast motion). At the same time, all diegetic sounds gradually fade out, to be replaced by non-diegetic, comforting music which serves as a counterpoint to the shots of night clubs and busy traffic. By doing so, film director Michael Winterbottom makes the audience's attention shift from the urban setting to Nadia's mood and what *she* needs while moving through the city. Winterbottom repeats this strategy in a scene in which a character named Eddie panics after receiving the news he will become a father. While Eddie leaves his London house to take a drive on his motorcycle all night, we again see fast motion and hear how all urban noises are filtered out except for the sound of the motorcycle. Eventually, even this noise recedes and is masked by music. Film theorist Barbara Flückiger has argued this is an effective way to signify that characters are not in touch with reality:

> The disappearance of noises marks or indicates loss of reality. The subject, the character, seems disconnected from the acoustic surroundings and, thereby, from reality.[53]

It is a use of sound that composer and professor of post-production Stephen Deutch has coined »Heightened FX«: it fuses »literal sounds« (sounds which »encourage us to believe what we see«) and »emotive sounds« (sounds which »encourage us to feel something about what we are seeing«) into a »single gesture«.[54]

Aside from this use of urban sound, it is a common strategy to contrast the sonic character of individual neighbourhoods within a city. Films set in pre-unification Berlin, for instance, tend to contrast the relative silence of the border zone or Potsdamer Platz in East-Berlin with the dazzling energy of West-Berlin – silence and natural sounds signifying the sonic absence of metropolitan life, or, for that matter, the need to be silent in order to hide oneself.[55] And Liselotte Engster's 1972 sonic collage of Berlin, entitled *Berlin – laut und leise* (Berlin – aloud and quiet) expresses the contrast between the tranquillity of a park idyll and the noises produced by airplanes, trams and cars. It would be similarly interesting to unravel how particular groups of migrants in particular urban neighbourhoods are portrayed in terms of sound. Jennifer Stoever-Ackerman has done so in her analysis of *New York Times* articles, published in the 1950s, about Puerto Rican migrants and of Tony Schwartz' sonic portrait of Midtown New York (*Nueva York: A Tape documentary of Puerto Rican New Yorkers,* 1955). While *The New York Times* consistently created a contrast between loud, boisterous and noisy Puerto Ricans and quiet white middle class people, Tony Schwartz' radio portraits were sonically far richer in their interpretation of both sides of the »sonic color-line«.[56]

If the strategy of juxtaposing different *locations* within a city has been used to stage and capture urban sound, the same has been done by juxtaposing different *time periods*. We can find examples in Annelies Jacobs' study of diaries kept by Amsterdam

53 Flückiger 2001, 397.

54 Deutch 2007, 4.

55 Aalbers 2012, forthcoming.

56 Stoever-Ackerman 2010.

residents during World War II. The sudden and dramatic change of the city's soundscape at the start of the war, with its intrusive sounds of airplanes, anti-aircraft gunfire and air-raid alarms, contributed to an atmosphere of fear and bewilderment, notably at night.[57] During the days and evenings, in contrast, the gradual disappearance of crowds and traffic – due to curfew, the confiscation of cars and bicycles, and the deportation of Jews – created an awkward silence. On May 18, 1940, one diary writer characterized Amsterdam as »very calm« and with »no traffic and no damage and very few Germans about«.[58] Another diarist wrote the following about the situation one week later:

> Virtually no cars anymore in the streets. The only transport still permitted is that of troops and groceries. City bus services have been discontinued or reduced. Doctors may no longer visit their patients by car.[59]

Fuel scarcity made private cars gradually vanish from the city's streets, while bicycle taxis were introduced to compensate for the loss. Similarly, horse-drawn carriages served as taxi. In 1940, after a visit to the Leidsche Plein Theater (»to forget our worries«), one diarist wrote:

> When we came out of it [...] it was pitch dark under the trees of the Leidsche Plein. With the silhouettes of people, horses and carriages (that now replace the taxis), and the weak lights of arc lamps, bicycles and trams, the scene had an old-fashioned ring to it – previous century or so.[60]

The sound of the occupied city reminded residents of the old days. It was hardly a welcome silence, though. If the silence reminded the writer of the sound of the *past*, it did so as a metaphorical shift in time forced upon the city residents – as a sign of retrogression, rather than in any pleasant *nostalgic* way. As two other diarists put it:

> It is ever more striking how few people go to work in the city between 8 and 9 o'clock. But the trams are overcrowded, with countless people standing on the

57 Jacobs 2013 forthcoming.

58 Archives *Nederlands Instituut voor Oorlogsdocumentatie* (NIOD) [Dutch Institute for War Documentation], no. 1082, Diary Edith Cox, May 18, 1940.

59 Archives NIOD, No. 1230, Diary C.J. van Buuren, May 25, 1940.

60 Archives NIOD, No. 1187, Diary Gerlof Verwey, September 5, 1940.

> footboards […]. The hundreds, or thousands, of bicycles are gone; inasmuch not confiscated by the German army, they are useless anyway for lack of tires.[61]
>
> The city is calm, without trams; there is much bicycling and the thousands of those on reduced pay use their leisure time by bicycling to or from Hoorn in order to try and find something to eat. This goes on despite the constant checks and confiscation of bicycles.[62]

In 1943, diarist Mirjam Levie expressed the eeriness of this silence and its »backward« character very explicitly:

> On Muntplein it is so quiet that there is no police anymore. You hardly see cars anymore in Amsterdam, only those of the German army and a few others. There are also fewer bicycles, and all this has made traffic so low that traffic police officers have absolutely become superfluous. […] But then the Jewish quarter, such as the Breestraat; it is deathly quiet there because many of the people were carried off. […]I wanted to tell you this because the city looks poor, deserted and dirty.[63]

In »Bombs«, a poem by Paul Rodenko written in Amsterdam during the war, it is not the evocation of the sounds of the *past* that expresses the negative experience of silence, but a reference to a modern spatial *future*. This is how Rodenko stages the threatening, fearful silence after the air-raid alarms have sounded:

De stad is stil.	The city is silent.
De straten	The streets
hebben zich verbreed.	have broadened.
Kangeroes kijken door de venstergaten.	Kangaroos are looking through the window gaps.
Een vrouw passeert.	A woman passes by.
De echo raapt gehaast	The echo hurriedly picks up
haar stappen op.	her footsteps.[64]

61 Archives NIOD, No. 1141, Diary C.A.L. Sander, February 4, 1943.

62 Archives NIOD, No. 1570, Diary Anton van Donkelaar, 30 October 1944.

63 Bolle 2003, 98.

64 Rodenko 1951, 14.

As critic Kees Fens suggested, the woman's footsteps »renew« the silence.[65] But what interests us even more here is the reference to the »broadened«, more spacious streets. Although the poet uses this phrase in a figurative sense here, designing wider streets was in fact one of the interventions fostered by noise abatement societies to combat the problem of city noise in the 1930s. Modernizing urban space by straitening and broadening streets, so was the idea, would enhance a smooth flow of traffic and reduce the reverberation of noise in narrow streets with tall buildings.[66] In Amsterdam, this had been realized in several of the newer neighbourhoods, but in the 1940s it was still largely a modernist planning dream. In Rodenko's poem, the reference to a more spacious urban space highlights the silence. Simultaneously it may also anticipate, just like windows without glass, the near future within the realm of the poem itself. As the next stanza puts it: »Quietly three four bombs land onto the square / and three four houses slowly run up / their red flag« (*Geruisloos vallen drie vier bommen op het plein / en drie vier huizen hijsen traag / hun rode vlag.*).[67] The streets were broadened indeed, also in a literal sense. The study by Jacobs of the Amsterdam war diaries equally demonstrates how contrasts between the interpretations of past, present and future sounds sharpened the experience of everyday life during the war. Sounds associated with the widespread problem of neighbourly noise in the pre-war years[68], for example, acquired new meanings during the war. City residents would listen in fearful anticipation for sounds that could mean bad news, such as the door bell, which was potentially a rather sinister sound, especially when rung at an odd time. In contrast, merely hearing familiar sounds from the house next-door was now often interpreted as a deeply comforting sound rather than a nuisance, for it meant that the neighbours were not yet arrested by the enemy. Such shifts in the meaning of sounds and modes of listening during the war provide us with new cues on how the war was experienced, as framed in part on the basis of earlier experiences. The diarists and the poet captured their sonic perceptions of a disrupting episode by comparing them with imaginary past and future soundscapes, just like radio and filmmakers sonically contrasted neighbourhoods to express their view on urban identities.

65 Fens 2007, 65.

66 Bijsterveld 2008.

67 Rodenko 1951, 14.

68 Bijsterveld 2003.

6. Getting Used to Sound

So far, we have shown how those writing about the urban soundscapes they inhabited – temporarily as a visitor or as a longtime resident – or those presenting fictional city soundscapes in radio plays and films often used narrative transitions in time and space to foreground sounds articulating the urban experience. A focus on such transition comes with a methodological risk, however. One might easily forget about less prominent ways of staging sounds in texts, plays and movies, and thus actually reproduce the striking aspects of urban sound: the iconic sensory overload of the urban environment in its sensational or intrusive aspects, or city sounds in their reassuring and sinister roles, or urban sounds as the markers of particular locations in the metropolis. To counter this risk, we also considered how authors and protagonists »talk about« or »display« adaptation to urban sound.

In *Sensing Changes*, Joy Parr has recently presented her research of several mid-twentieth-century mega projects in Canada – such as the building of a military base, a nuclear power plant and a dam – and »the processes by which inhabitants adapted to the habitats the megaprojects had transformed«.[69]

> As inhabitants incorporated into their bodies the altered world beyond their skins, awarenesses they usually held beyond telling, as habit and reflex, became urgently speakable. In *Sensing Changes*, we encounter embodiment both as active adaptation to changed circumstances and as »the whispering of ghosts«, relicts of past successful adaptations to familiar worlds later remade, persisting as familiars, reminders of losses, and also sources of resilience and resources for rebuilding.[70]

In »Moving and Sound«, one of the chapters in *Sensing Changes*, Parr tells how the Ontario village Iroquois and the St. Lawrence River were remade as a consequence of the establishment of a big dam in the late 1950s. She shows that bodily techniques, which villagers had developed in their daily interaction with the old river and village, were of no use anymore in the newly constructed, barren districts. Walking became less common, gathering together around cars in front of shops became a mere memory, and the sounds of the river – the throbbing engines of ships, the slapping sounds that signalled rough days – did no longer make up the keynote of village life. The villagers' bodies had been »familiars of the river«, but these bodies turned lonely with the arrival of the dam,

69 Parr 2010, 1.

70 Parr 2010, 2.

transforming the sensory knowledge of their former surroundings into the »embodied lostscapes« they became aware of once under threat.[71]

Again, we see how transitions made people aware of their everyday sensory experience of their environment, and how memories of these changes and of the way in which they dealt with them spurred them to make these experiences »speakable«, in the same way as breaches of conventions render the taken-for-granted visible. Yet how did historical actors, radio play writers and filmmakers express the phenomenon of adapting to sound, of simply accepting the sound of the urban environment as it was?

Jacobs' study of diaries by Amsterdam residents during World War II again provides several cues. As clarified above, the sounds of airplanes, anti-aircraft gunfire and air-raid alarms initially caused bewilderment among the Amsterdam population, most of whom had not experienced such sounds before. Yet the diaries also show how some people soon got used to these sounds, while others did not. On 20 May, C.J. van Buuren wrote that his wife still heard the airplanes at night, but that he managed to have a good night's sleep again. Only by the end of the month the same was true of his wife. In June he occasionally heard shootings, at times heavy or nearby, and later at a large distance. He no longer recorded the sounds he heard, nor did he write about airplanes; he merely referred to »activity in the air«. Over time, his choice of words started to underscore the repetitive nature of the air fights: »At night a constant drone of airplanes. Gunfire at a far distance« (9 July); »a few air strikes with accompanying noise« (24 July); and: »Next, the usual noise of airplane drone and gunfire; after half an hour, a repeat of the whole episode« (27 July).[72]

Whenever diarists experienced the war violence intensely, for instance because it came close, they paid much attention to sound. Airplanes proved to »roar« and »drone«, the alarm »howled« and the anti-aircraft gunfire »pounded«. Together these sources of sound would »boom«, particularly when they came so close that the sound could be felt or the house started quivering:

> This is the most awful night we have yet experienced. It has been so bad, so horrible that for the first time I was really scared, which doesn't happen easily. At half past 2 the 1st heavy bombs; directly followed by pounding fire, so heavy that we were shaking in bed. It lasted forever and its intensity kept increasing. [...] Shell fragments dropped into the street; everything seemed to take place right above our heads. In between the anti-aircraft gunfire one could vaguely hear the drone of airplanes. Whether there were any fights in the skies over the city we don't know.

71 Parr 2010, 87, 3.

72 Archives NIOD, No. 1230, Diary C.J. van Buuren.

> You could not come near the windows because of the risk. All was booming and banging. It was nerve-racking, and it lasted until after 4, without interruption.[73]

Yet adult diarists seemed to get used to the sounds of war later on, even if they continued to be aware of the danger:

> At 20.15 the gunfire aimed at airplanes coming over started, and it went on, with some brief interruptions, until 22.15. Two hours of booming and rumbling, but we hardly listen to it anymore. The radio plays on and we read a book, hardly noticing what goes on outside.[74]

Air-raid alarms, airplanes and gunfire were now referred to with metaphors such as »show«, »concert«, or »party«. As one diarist put it, one did not contribute to such »party«, to be sure, by falling asleep.[75] Furthermore, several diarists indicated that the sound of airplanes and alarm also prompted joy, because all the air activity suggested that allied forces were back again to fight the Germans:

> When there is an air-raid alarm, you hear cheering in the street right away: fortunately, they are back again!! No one is afraid. They won't get us, those krauts.[76]

The quotes thus display both sensory stress and processes of accommodation – as some people gradually learned to sleep through the anti-air raid sirens.
The diaries also show evidence, and heartbreakingly so, that children never quite adapted to the noise of war, perhaps because it was harder to them to give meaning to it or put it into perspective. Anne Frank, for instance, was almost eleven years old when the war broke out. She repeatedly indicated to be afraid of what she mostly referred to as »shooting«; in nights with shootings she would look for comfort at her father's side.[77] On 2 June 1944, when she and her relatives had been in hiding for nearly two years, she invented »a brand new anti-shooting recipe«.[78] It basically implied that in case of »hard poppers« one should run up and down a wooden stairs and, preferably also fall

73 Archives NIOD, No. 1151, Diary C.M.A. Bruijn-Barends, 6 July 1940.

74 Archives NIOD, No. 1230, Diary J.C.van Buuren, 27 January 1942.

75 Archives NIOD, No. 1230, Diary J.C.van Buuren, 27 March 1943.

76 Archives NIOD, No. 1151, Diary C.M.A. Bruijn-Barends, 23 June 1943.

77 Frank 2008 [1942-1944], 112.

78 Frank 2008 [1942-1944], 274.

down cautiously a few times – just so one did not have to think about the shooting for a moment. The noise thus generated indoors reflected a child's imaginative effort to mask the dreadful noise of the shooting outside. Likewise, Gerlof Verwey underscored in his diary how his children continued to be afraid of the sounds of war, while he himself had gotten used to them, or, at least, that is what he pretended:

> Throughout the night a howling of the sirens and shooting – baby Ger crying all the time, Abje shaking in his bed, William with eyes open – the whole family awake. [...] 8.30. Bombs falling, the siren howls. The gunfire is pounding. We get used to it. Strikingly we continue to work calmly, pretending that nothing is going on. A sense of fatality has grown stronger.[79]

Our other sources similarly display the complexity and subtlety in processes of adaptation to sound that had once been new. In his study of films, Jasper Aalbers found that adaptation to urban sound was not only expressed through storylines, as in Out of the Clouds on the German and American tourists in London, but also in changes in the character of the staged sounds and in the relationship between sounds onscreen and off-screen. An example is the remarkable role of traffic sound in films set in Berlin. A comparison of East-Berlin films made in the 1950s with those produced in the 1970s reveals that the relative presence of *unspecified* background sounds had increased: from 9.4 percent of all »urban sound events« (counted by Aalbers) in the 1950s films to 27.4 percent in the 1970s. In contrast, the relative presence of *distinct* traffic sounds decreased, from 45 percent in the 1950s to 34.1 percent in the 1970s, even though the share of distinct *car* sounds within these traffic sounds had slightly risen. As the actual increase of cars in the GDR had been much higher, going from 75,710 cars in 1950 to 1,159,778 in 1970[80], the staging of sound in the East-Berlin films seemed to present car sound as a nearly naturalized aspect of the key sounds of East-Berlin, and thus, in Aalbers' view, as something people had become used to. However, while distinct car sounds were onscreen in 70 percent of the cases in his Berlin sample of the 1970s, this was even 89 percent in his East-Berlin sample. This particular staging of sound, Aalbers argues, shows that the presence of cars had a less taken for granted status in his East-Berlin sample than in his Berlin sample at large, thus expressing the relative novelty of the car's predominance in East-Berlin after all. Even if the sound of cars increasingly coloured the keynote of East-Berlin, films displayed it more often on screen, thus presenting it as relatively »special«.

79 Archives NIOD, No. 1187, Diary Gerlof Verwey, 15 November 1940.

80 Deutsche Demokratische Republik 1975, 245.

7. Conclusions

At the end of the 1970s, the well-known Dutch columnist Henk Hofland acknowledged that he loved to read about the »sounds of yesterday« because they made him feel deeply »nostalgic«. Recalling the sounds of the past gave him a »sense of unspeakable approval of life« similar to what he had felt as a young kid, lying in bed, and listening to an early morning horse tram driving into town.[81] His effusion reminds *us* of Rutger Zuyderfelt's »Take a closer listen«, discussed in our previous chapter, which radiates the same thirst for urban life through sound. Hofland, however, added a disclaimer – one which historians of sound are confronted with more often – suggesting that »a more or less systematic exploration« of this issue would show that such nostalgia for the sounds of yesterday was »nice but of little significance«.[82]

Without discrediting the role of sounds in eliciting nostalgia, this chapter has aimed to show that studying the sounds of the urban past and their media staging is relevant for presenting and articulating experiences of city life. A focus on sound, in other words, provides a fresh angle on the constructed identities of cities and their residents. Our argument reveals that authors and makers of texts, radio plays and films have deployed well-known topoi of urban sound, such as intrusive and sensational sound. We have also shown how across different media the same narrative tools were employed for capturing and staging sound: the urban arrival scene, showing the daily rhythm of urban life, juxtaposing the soundscapes of different neighbourhoods within one city, and contrasting the soundscapes of the present, past and future. As our argument demonstrates, these strategies have been deployed across the entire time period under study, from the late nineteenth until the early twenty-first century. At the same time, however, we elucidated the historicity of the wording or sounds that figure in these narrative conventions. For example, while the musicalization of sonic vocabularies was typical for works from the late nineteenth century, the noisification of music was characteristic for, though not fully confined to, works from the early twentieth century.

Moreover, in addition to conventions we tracked down less predictable representations of urban sound, such as the silence and silent episodes in the rhythm of urban life, the staging of adjustment to urban sound or the remarkable iconization of *keynote* sounds like the sounds of children playing outside, which may also be coined as the acoustic profiling of particular neighbourhoods. This means that the texts, plays and films we studied did not only reify existing representations of the city, but also opened up

81 Hofland 2010 [1978], 76.

82 Ibid.

our ears to less traditional ones. This means that studying the representation of sound in media products of the past involves much more than reconstructing »lostscapes«, the past sensory experiences of vanished places, as Joy Parr aimed to do in *Sensing Changes*. Such effort is as much about the sounds and experiences »won«, like the sound of traffic on smooth asphalt or the wonderful sounds of neighbours still around in war-ridden Amsterdam of the early 1940s (instead of them having been arrested and removed by the enemy), as it is about sounds and experiences lost.

As regards the study of auditory topoi of mechanical sounds in literature, the basic concern in *Mechanical Sound* (see Chapter One) was to acquire a better understanding of how lay people dramatized sound when complaining of noise, and to understand why their ways of doing this often clashed with the ways of dramatizing sound required in formal legal settings, such as hearings in the context of nuisance law. In this chapter, however, we studied how writers, radio play writers and filmmakers used auditory topoi and other tools to open up *our* ears, thus forcing us take a closer listen where usually we refrain from doing so. It is in the *ways* in which they made us listen that we may find experiment and novelty. Capturing and staging the experience of urban sound may not automatically lead to convincing artistic renewal, as the critique of Russolo's art noise has revealed. Yet at times, struggling with capturing sound does lead to unexpected innovations.

Writers and the makers of radio plays and films have found new ways to stage the experience of urban sound: fleeting, fugitive, fragile and evanescent, like so many other sensory experiences. They did so by blending everyday sounds with music, or amplifying music at the expense of city sounds, or playing the background sounds of traffic while varying the extent to which they were seen onscreen and off-screen, to mention just a few examples. If we were looking for conventions, we harvested the unexpected in the same act. One type of intermediality, as will explain in more detail in Chapter Four, is taking up the formal characteristics of a particular medium within another, different medium.[83] The musicalization of noise and noisification of music are similar examples of mixing the characteristics of two »genres« in order to evoke a particular experience of the city.

Such innovations can be highly informative for museum curators as well. We started our chapter with the claim that the use of natural sound and sound recordings can be an interesting solution to the authenticity problem historical museums struggle with – bringing audiences close to past objects without enabling them to touch these objects – provided that presenting sound and explaining the complexities of its historical meanings go hand in hand. *Sound Ways*, the Finnish exhibition we opened this chapter with,

83 Rajewsky 2005.

did so by combining sound and text, and by making use of a literary strategy: employing the audiograph. Other recent historical exhibitions and historical soundwalks combine sounds and sound effects with recorded narratives drawn from oral history interviews, thus adding context *and* authenticity through the reports of earwitnesses.[84] The cultural heritage of radio, film and text and the strategies of their makers for capturing sonic experiences can add interestingly to this palette. First, museum curators may well deploy the cultural heritage of radio, film and text as illustrations. *Weekend* may then serve as an example of a particular, rhythmic staging of experiencing the city in the 1930s.[85] Perhaps these curators will also find inspiration in the narrative strategies used by writers and makers of radio and film to capture the varied auditory experiences of the city. This especially helps them to cater those visitors who »hope to hear imaginative narrations playing into the gallery, mixed with beautifully crafted, ambient sound«.[86] Having museum visitors experience the sounds of the urban past through the stories of past visitors is one option, but one could also invite museum visitors to record their own initial sonic surprises when *arriving* in the city in which the museum is located, or ask them for their memories of such first encounters.[87] Moreover, a focus on the changing words used for sounds – for instance by creating compositions of such words – may give visitors an idea of vanished sounds (some words will refer to obsolete artefacts or practices) and of changing ways of evaluating such sounds, like the assessment of sound in terms of musicality versus acoustics. Staging the topoi of the intrusive, sensational, sinister and comforting sound with historical examples – such as fear of the sounds of wild animals in a lonely countryside versus the soothing buzz of cities announcing the presence of fellow citizens – will also help museum visitors to understand past experiences.

In addition, a focus on urban *rhythm* may allow one to evoke a day in the life of a city at some point in the past and provide a richer sound-image of urban life than the stereotype of the pastoral, quiet pre-industrial past versus the noisy post-industrial present. And, sonic *juxtaposition*, of areas or neighbourhoods within a city for instance, may make museum visitors aware of the differences in social conditions within specific urban settings. Over the years, authors, filmmakers and writers of radio plays have found various ways to express how people managed to adapt to and cope with shifting urban sounds, thus highlighting their meanings to different groups in shifting political and

84 See, for instance, Bryan / Vyner 2012 and Hutchison / Collins 2009.

85 See Technische Sammlungen Dresden.

86 Angliss 2005, 20.

87 The Memory Machine, an installation at the British Museum, recorded and played back oral memories provided by visitors about the British Museum and its collections. See Lane / Parry 2003, 9.

socio-economic circumstances. By presenting such rich and original stagings of urban sound in exhibitions, museums can show that Hofland's nostalgia for the comforting sounds of his childhood represents only one way of listening to the past.

8. Sources

Aalbers, Jasper (2012 forthcoming): Echoes of the City. Staging Urban Soundscapes in Fiction Film, 1930-2000, Maastricht.

Aboa Vetus & Ars Nova Museum http://www.aboavetusartsnova.fi/en/nattelyt/sound-ways (Accessed July 24, 2012).

Angliss, Sarah (2005): Sound and Vision, in: The Museum Journal 105.7, 26-29.

Anonymous (1878): Brieven uit Amerika, New York, 20 April: De Phonograaf, in: Het nieuws van den dag. Kleine courant, May 21 (Tweede Blad), 1.

Anonymous (1921): Futuristische muziek, in: Het Vaderland, June 20 (Avondblad), B2.

Barr, Charles (1999): English Hitchcock, Moffat.

Bijsterveld, Karin (2003): »The City of Din«. Decibels, Noise and Neighbors in the Netherlands, 1910-1980, in: Osiris 18, 173-93.

Bijsterveld, Karin (2008): Mechanical Sound. Technology, Culture and Public Problems of Noise in the Twentieth Century, Cambridge MA.

Bolle, Mirjam (2003): Ik zal je beschrijven hoe een dag er hier uitziet. Dagboekbrieven uit Amsterdam, Westerbork en Bergen-Belsen, Amsterdam.

Braunger, Amely (2008): Intermedialität und Urbanität im Medienwandel. Eine Untersuchung zur Darstellung der urbanen Lebenswelt in Hörspiel, Roman und Film am Beispiel von »Berlin Alexanderplatz«, Konstanz.

Brunsdon, Charlotte (2007): London in Cinema. The Cinematic City Since 1945, London.

Bryan, James / Vyner, James (2012): Soundmap. Bringing History to Life Through Audio Tours. Unpublished presentation, Symposium Staging Sound in the Museum: the Ears-on Experience of History, February 3, 2012, Hilversum, Institute for Image and Sound, The Netherlands.

Burstyn, Shai (1997): In Quest of the Period Ear, in: Early Music 25.4, 693-697.

Coster, Charles Théodore Henri de / Janson, Karel J. A. (1998 [ca. 1879]). Zijn vrouw zet bloedzuigers. De zeer persoonlijke indrukken van een reis, in 1877 ondernomen naar Amsterdam, Zaandam, Broek in Waterland, Monnikendam, Volendam, Hoorn, Scheveningen, Marken en Urk (3rd ed.), Amsterdam.

De Amicis, Edmondo / Lodeesen, D. / Fagel, Roland (1985 [1876]): Nederland en zijn bewoners, Utrecht.

Deutsch, Stephen (2007): Editorial, in: The Soundtrack 1.1, 3-13.

Deutsche Demokratische Republik (1975): Statistisches Jahrbuch der Deutschen Demokratischen Republik 1974, Berlin.

Fens, Kees (2007). Stilte. in: Fens, Kees, In het voorbijgaan. Kleine essays, Amsterdam, 64-66.

Fickers, Andreas (2012): In Search of the Sonic Aura. Challenging the WYSIWYG Approach in Historical Exhibition Making. Unpublished presentation, Symposium Staging Sound in the Museum: the Ears-on Experience of History, February 3, 2012, Hilversum, Institute for Image and Sound, The Netherlands.

Flückiger, Barbara (2001): Sound Design. Die virtuelle Klangwelt des Films, Marburg.

Frank, Anne / Frank, Otto (2008): Het Achterhuis. Dagboekbrieven 12 juni 1942-1 augustus 1944, Amsterdam.

Goergen, Jeanpaul (ed.) (1989): Walter Ruttmann. Eine Dokumentation, Berlin.

Gouw, Jan ter (1974 [1875]): Amsterdam en de Amsterdammers door een Amsterdammer, Amsterdam.

Hake, Sabine (2004): Urban Paranoia in Alfred Döblin's Berlin Alexanderplatz, in: The German Quarterly 67.3, 347-68.

Havard, Henry (1876). Amsterdam et Venise, Paris.

Haus der Musik (Austria): Catch The Sound, http://www.hausdermusik.at/en/sound-museum/16.htm, 2012/07/24.

Hofland, Henk J.A. (2010 [1978]): Oud geluid, in: Hofland, Henk J.A., De kronieken van S. Montag: Nederland 1975-2010, Amsterdam, 75-77.

Howes, David (2011): Vienna. Sensory Capital, in: Diaconu, Mădălina / Heuberger, Eva / Mateus-Berr, Ruth / Vosicky, Lucas Marcel (eds.), Senses and the City. An Interdisciplinary Approach to Urban Sensescapes, Vienna, 63-75.

Hutchison, Mary / Collins, Lea (2009): Translations. Experiments in Dialogic Representation of Cultural Diversity in Three Museum Sound Installations. In: Museum and Society 7.7, 92-102.

In Flanders Fields Museum (Belgium): http://www.inflandersfields.be, 2012/07/24.

Jacobs, Annelies (2012 forthcoming): Barking and Blaring. City Sounds in Wartime, in: Biddle, Ian / Gibson, Kirsten (eds.), Noise, Audition, Aurality. Histories of the Sonic World(s) of Europe, 1500-1945, Oxford.

Jacobs, Annelies (2013 forthcoming): Het geluid van gisteren. Waarom Amsterdam vroeger ook niet stil was, Maastricht.

Jelavich, Peter (2006): Berlin Alexanderplatz. Radio, Film, and the Death of Weimar Culture, Berkeley.

Lane, Cathy / Parry, Nye (2003): The Memory Machine. Sound and Memory at the British Museum, Paris.

Lehto-Vahtera, Johanna (2010): Ääniteitä. Näyttelyn aineettomuus ja museologia (Sound Ways. Immateriality and Museology), in: Hannula, Linda / Lehto-Vahtera, Johanna (eds.), Moniääninen, Turku, 10-16.

Letschke, Rainer (2003): Einführung in die Medientheorie, Munich.

Marsh, Caryl (1998): Exhibition Review. In Praise of Sound at the Royal Ontario Museum, in: Curator 41.1, 50-56.

Mason, Jennifer / Davies, Katherine (2009): Coming to our senses? A critical approach to sensory methodology, in: Qualitative Research 9.5, 587-603.

Maurik, Justus van (±1890): Amsterdam bij dag en nacht. Novellen en schetsen, Amsterdam.

Muhonen, Timo (2010): Ääni näyttelyn fokuksessa. Rajoitteita ja ratkaisuja (Focus on Sound in an Exhibition. Limitations and Solutions), in: Hannula, Linda / Lehto-Vahtera, Johanna (eds.), Moniääninen, Turku, 17-27.

Museum Het Domain (Netherlands): Sound Souvenirs / Tapemosphere 2007-08, http://www.hetdomein.nl/index.php?p=308&t=content, 2012/07/24.

Ortigão, Ramalho / Jong, Marcus de (1948 [1885]): Holland 1883. Amsterdam.

Parr, Joy (2010): Sensing Changes, Vancouver / Toronto.

Rajewsky, Irina O. (2005): Intermediality, Intertextuality, and Remediation. A Literary Perspective, in: Intermédialités 6 Automne, 43-64.

Rehhahn, Yvonne (1996): Nur der veranderte Autor kann den Film verandern, in: Berlin Alexanderplatz. Drehbuch von Alfred Döblin und Hans Wilhelm zu Phil Jutzi's Film von 1931, Munich.

Rieger, Matthias (2003): Musik im Zeitalter von Sound. Wie Hermann von Helmholtz eine neue Ära begründete, in: Phleps, Thomas / von Appen, Ralf (eds.), Pop Sounds. Klangtexturen in der Pop- und Rockmusik. Basics – Stories – Tracks, Bielefeld, 183-96.

Rodenko, Paul (1951): Bommen. In: Gedichten, Amsterdam, 14.

Ruhrmuseum (Germany): http://www.ruhrmuseum.de/museum/, 2012/07/24.

Ruttmann, Walter / DJ Spooky et al. (1998): Weekend Remix, http://www.medienkunstnetz.de/werke/weekend-remix/, 2012/07/31.

Shapins, Jesse (2008): Walter Ruttmann's *Weekend*. Sound, Space and the Multiple Senses of an Urban Documentary Imagination, http://www.jesseshapins.net/writings/JShapins_RuttmannWeekend_Jan2008.pdf (Accessed July 27, 2012).

Smith, Mark M. (2012): The Garden in the Machine. Listening to Early American Industrialization, in: Pinch, Trevor / Bijsterveld, Karin (eds.), The Oxford Handbook of Sound Studies, Oxford, 39-57.

Sterne, Jonathan (2003): The Audible Past. Cultural Origins of Sound Reproduction, Durham.

Steward, Jill / Cowan, Alexander (2007): Introduction, in: Cowan, Alexander / Steward, Jill (eds.), The City and the Senses. Urban Culture Since 1500, Aldershot, 1-22.

Stocker, Michael (1994): Exhibit Sound Design for Public Presentation Spaces, in: Museum Management and Curatorship 13, 177-83.
Stoever-Ackerman, Jennifer (2010): Splicing the Sonic Color-Line. Tony Schwartz Remixes Postwar Nueva York, in: Social Text 28.1, 59-85.
Technische Sammlungen Dresden: http://www.tsd.de/sammlungen.
Thompson, Emily (2002): The Soundscape of Modernity. Architectural Acoustics 1900-1933, Cambridge MA.
Timmerman, Aegidus Willem (1938): Tim's herinneringen, Amsterdam.
Titon, Jeff T. (2008): The paradox of authenticity; http://sustainablemusic.blogspot.nl/2008/09/paradox-of-authenticity.html, 2012/07/05.
Vogt, Guntram (2001): Die Stadt im Kino. Deutsche Spielfilme 1900-2000, Marburg.
Wood, Charles W. / Baarslag, C. (2002 [1878]): Door Nederland, Rijswijk.
Yasko, James (2006): Podcasting in Museums, in: Journal of Museum Studies 1.1, 1-35.

Why Historians of the Auditory Urban Past Might Consider Getting Their Ears Wet

Mark M. Smith

1. Introduction

A city might hug it, sit precariously above it, be in danger of being deluged by it, slip gently into it, or may be veined by it. Regardless of its particular location and specific configuration, the urban area that does not touch – and is not touched by – a body of water was, until the mid-twentieth century, quite a rare thing. As Lewis Mumford memorably showed in his *The City in History: Its Origins, Its Transformations, and Its Prospects*, cities, with relatively few exceptions, have nearly always been built on coasts, inlets, bays, harbours or along rivers. And for good reason: water allowed transportation of goods and people, offered sources of food and protection, and, during the industrial revolution, helped power machinery and factories. Bodies of water and urban environments became inseparable.[1] Can we really entertain a London without a Thames, a New York City without a harbour, a San Francisco without a bay, an Amsterdam without a canal system? No. Nor should we. But we do, at least when it comes to listening to the past. For the most part, we listen to, and examine, study, and make sense of the sound produced on land, physically in the city, and in so doing inadvertently muffle the sounds coming from the water. When it comes to aquatic soundscapes and how they functioned to help constitute urban soundscapes, we are, for the most part, quite deaf.

And unnecessarily so. To say that many cities are within view of the water is to also to say that many of them are touched by it and, for our purposes here, to also say that city and urban soundscapes very often have been constituted not simply by what sounds and noises were produced and heard on land but also by those at sea, on lakes or on

1 Mumford 1961.

(or, indeed, in) rivers. It might well be the case that most histories of urban soundscapes cannot profitably and reliably convey the full range and meaning of their subject without also listening to water. This is not to say that current work is interpretively at sea but it is to say that we might begin listening more attentively to the auditory interplay between sounds made by and on water(s) and how they interacted with – exporting and importing – the sounds of the cities. While those aquatic and terrestrial soundscapes were not necessarily always one and the same they cannot properly be understood without mutual reference.

2. Barely Audible Trickles. Historical Writing about Sound and Water

That point, of course, is quite obvious upon reflection but it is not one that has been made often or explicitly by scholars of acoustemology (a term Steven Feld helpfully used to capture the ways hearing and listening constitute knowledge of the world).[2] This is hardly the place for a full review of the historiographical holes; neither is it my desire to point to particular works and flag their eliding of aquatic soundscapes. If it were, the limitations of my own early work on nineteenth-century U.S. soundscapes would figure prominently. It is, though, worth saying that historiographical niches and subfields that one might expect to yield a sustained call or treatment of the acoustemology of seas, rivers and oceans, say very little about the matter. Certainly, we know a little about the sounds of certain rivers and certain types of water – waterfalls, especially – and we are often told what the water body might have looked like. But even as we keep water in the picture we oddly keep the sound of water out of historical acoustemology generally, the history of the urban soundscape specifically.[3]

In fact, the only sustained, in-depth treatment of the sound of water of which I know comes not from environmental or social historians – who one might reasonably think most interested in the topic – but, rather, from an art historian: the redoubtable Douglas Kahn. Kahn's *Noise, Water, Meat: A History of Sound in the Arts* offers a magnificent treatment of »the sound of actual water« within the late modernist arts' movement in the United States. Indeed, it might not be too much to claim that we probably know more about how people under and in the seas and oceans listened – courtesy of Hillel

2 Cf. Feld 1982.

3 On the sound of the Rhine, Erlmann 2010, 151-52, 165-66; on waterfalls, see Schwartz 2011, 37, 148, 261, 287, 362.

Schwartz's recent work on twentieth-century sonar technology used by submarines – than we do about how the sounds of the sea influenced experience and understanding by listeners on land and in cities.[4]

My inspiration for this brief essay – more properly, rumination – comes, then, not so much from existing literature but, rather, from two quarters. The first – and most immediate – source of inspiration comes from the tantalizing bits of evidence embedded in the essay »Shifting Sounds: Textualization and Dramatization of Urban Soundscapes«, by Karin Bijsterveld, Annelies Jacobs, Jasper Aalbers and Andreas Fickers (this volume). Oceans, seas, and rivers do not drive their essay but they do inhabit it in interesting, potentially useful ways and I would like, before expanding on their evidence, call attention to it.

In their essay, we have various references to the sounds of ship foghorns at sea, ship horns in East London, the putatively quiet canals of Amsterdam reminding people »of the far away murmur of the sea«, and, courtesy of the pioneering work of Joy Parr, hints of how the riparian sounds of the St. Lawrence River – throbbing ship engines, the slapping sounds of the river – evaporated with the building of dams and levees.[5] What's so refreshing – and useful – about these pieces of evidence is how they function to suggest the ways in which water informed urban soundscapes. Water not only possessed its own soundmarks that people in cities listened to – what we might consider water's natural sounds (its slapping, its association with quiet) – but the seas and rivers also served as aquatic platforms for various technologies that produced their own sounds, thus enabling us to historicise how water – and what it carried – sounded at various points in time. For example – and here I am pivoting beyond the aforementioned evidentiary nuggets a little – the sound of the Thames changed as maritime technology evolved. When steam replaced sail, the auditory texture of how the river sounded changed. So too did the Thames' auditory reach. Steam whistles likely penetrated much farther and deeper into the city than did the flapping of canvass and jostling of rigging associated with sail-powered shipping.

4 Kahn 1999, 242-59, quotation on 244; Schwartz 2011, 713-21. My own work (Smith 2001) says little about the matter and mentions water only in passing. One might expect historians of the environment to say a bit more but water does not figure very much at all in Peter A. Coates' recent, exploratory and thoughtful essay (Coates 2005). Similarly, the rivers, lakes and waterfalls veining the United States pop up rarely, even in work that traces human movement across the land. See, for example, Keyes 2009. And two recent, unusually thoughtful overviews of historical acoustemology do not make dedicated calls for water's inclusion in the future writing on sound. See Rath 2008, Rosenfeld 2011. But do see the attention to aquatic sounds by Parr (2010), noted below.

5 Parr 2010, 87.

3. Watery Sounds. Nature and Industrialization

My commentary here is informed by my own recent work on nineteenth-century soundscapes in the United States, specifically on early industrialization in the United States and also on the acoustemology of the American Civil War.
Contemporaries, at least in the nineteenth century, believed water relevant for their understanding of soundscapes and environment generally probably because water was so prevalent that its sounds were inextricable to many peoples' daily experiences. For example, in 1856, the American popular magazine, *Harper's New Monthly Magazine*, ran a brief story on the meaning of the sense of hearing. The piece did not highlight the sound of water; nor did it privilege the sound of land. Rather, the magazine writer offered a sense of how the sounds of water, land and air braided to create a sense of being in, and experiencing, the world. Sounds »are heard everywhere«, noted the writer. The writer urged readers to listen, for example, to the sounds of nature:

> We may step into the tearful landscape [of nature] on a spring morning, and join in the jubilant songs of early birds; we may throw ourselves into the waves, and shout for joy amidst the thunder of the ocean, or we may listen on the sandy sea-shore to the throbbing of his great pulse, as he rises from the vast deep and embraces the land with a stormy, long-drawn kiss.[6]

A touch overwrought and clearly a product of Romantic sensibility though the commentary is, the writer nevertheless grants us access to a nineteenth-century American sensibility that refused to segregate sounds and thereby reveals a habit of listening and a way of hearing the world that braided sounds on the land with sounds from and in the sea.
In my examination of early industrialization in the United States – located in New England during the first half of the nineteenth century – it became clear to me that how that process sounded was very much tied to water, specifically riparian forms. Early textile factories, such as the ones in Lowell Massachusetts, were powered by water, with rivers often inextricable to the very architecture of the factory. As one writer explained in 1843:

> The city of Lowell stands upon the Merrimack river; upon a point of land, formed by the Concord river, at its confluence with the Merrimack, and a bend in that

6 Anonymous 1856, 640.

> river, from which its direction is at a right angle with its former course. It is intersected by many canals, the principal of which is the Pawtucket . . . thus forming an island of the city; it being *entirely surrounded by water*.[7]

It was precisely this riparian-driven technology that helped the first generation of factory workers make the intellectual transition necessary for work in the loud, noisy, sometimes deafening textile mills. These earliest factory workers, mostly young women from the surrounding countryside, came to places such as Lowell with the sounds of water – rivers, seas, waterfalls – fresh in their minds. As historian Anthony F. C. Wallace writes of an early textile factory in Rockdale in the 1850s: »The mills themselves, powered only by water, whispered and grunted softly«.[8] But even when the mills were loud, bellowing, and cacophonous, workers framed what they heard, in part at least, within a pastoral idiom that often drew on and from water. »Directly below my window passes the combination of nature, and human invention, forming a canal«, wrote one Lowell worker in 1842. Water powered this »American Manchester«, and the traditional sound of rushing torrents could be heard literally in the mills.[9] Others explained that the noise was tolerable because Nature's volume could be similar. As one Lowell worker, Susan, put it:

> You know that people learn to sleep with the thunder of Niagara in their ears, and a cotton mill is no worse, though you wonder that we do not have to hold our breath in such a noise.

Susan reached into the loudest sound of rural New England – the roar of Niagara – and made sense of her new aural environment on the factory floor.[10] This first generation of industrial workers interpreted the sounds of the urban, industrial environment in terms of natural sounds they already knew, helping them to not only make the transition to an urbanizing, industrializing society but also allowing them to calibrate changes in the urban environment thereafter.

7 H.F. 1843, 145, 147; Lewis 1844, 242. For a fuller discussion, see Smith 2012, 39-57.

8 Wallace 1978, 4.

9 M.T. 1842, 57.

10 Eisler 1998, 52, 51.

4. Shhh, Bang. Listening to the American Civil War

Listening to particular events during the American Civil War (1861-1865) is also instructive, revealing how a fuller understanding of how participants experienced and understood the war cannot rely solely on what they heard on land and in cities. Take, for example, the opening salvo of the war in April, 1861. That took place in Charleston, South Carolina, at the time one of the largest cities in the United States. While most people – Confederate soldiers and civilians – were located in the city, they listened actively to the noises, sounds, and silences emanating from the watery harbour that arced around the city. Nestled in Charleston's expansive, Atlantic-fed bay was a Union garrison, Fort Sumter, a man–made granite island that had just been captured by federal, Union forces and was now, at the beginning of the war, in their possession. Heavy guns from Charleston were directed at Fort Sumter in an effort to dislodge the Union troops. But, equally, Fort Sumter fired back, the exchange of booming of cannon between city and fort reverberating around the harbour. Sounds of martial might and noises of war skipped off waves and rattled around the bay, giving the entire battle a heightened auditory feel, a sense of unprecedented loudness that impressed contemporaries deeply. This was a bloody war that started with exceptionally loud bangs, thanks in part to the water over which shot was fired.

Moreover, the federal troops who initially captured Fort Sumter – on the night of December 26, 1860 – did so by paying attention to what was happening on the water. To capture the fort, Union troops, under the command of Major Robert Anderson, had to row several boats at night from the other side of the harbour (they were stationed at Fort Moultrie, opposite Charleston) to Fort Sumter. To execute the stealthy manoeuvre successfully, they listened intently for the presence of Confederate steam-powered boats patrolling the harbour and went to lengths to muffle their own sounds so that their voices and sounds of their oars did not travel over the Bay and alert Confederates in and around Charleston of their movements. My point is this: if we listened simply to the sounds of Charleston the city at the beginning of the American Civil War we would be in possession of a warped and impoverished story, one that would could not satisfactorily explain how Fort Sumter was captured by Union forces or fully describe why the battle had such resonance – literally – for the soldiers at the fort as well as for those people in the city. By taking the body of water around the city into account, we are in possession of a more textured and, dare I say, accurate account of the event, one that is acoustically inclusive as well as explanative.[11]

11 These observations are based on my ongoing work, Smith (forthcoming).

5. Conclusions

Until we start reading the auditory history of, for example, the Thames into the auditory history of the city of London, the history of bays, harbours and seas into the history of urban environments of which they were arguably a part, and the history of rivers into the history of early industrialism, we possess interpretations that unduly favour the soundscape of *terra firma*. Not only is this unfortunate, it is also misleading simply because for the people who lived and worked in those urban environments the presence of water mattered not least because it influenced how they conducted their trade, moved their goods, exchanged information, harvested food, fought wars and generally understood their environment. All of these activities constituted in part what it meant to be a city dweller and all of them had their auditory meaning and soundmarks. If we are to attend to the sensate past fully, we should be careful to sound not just the city but, when appropriate, sound the bodies of water that often formed part and parcel of those places.

6. Sources

Anonymous (1856): The Senses. IV. Hearing, in: Harper's New Monthly Magazine, New York, 634-41.

Coates, Peter A. (2005): The Strange Stillness of the Past. Toward an Environmental History of Sound and Noise, in: Environmental History 10, 1-33.

Eisler, Benita (1998): The Lowell Offering. Writings by New England Mill Women, New York.

Erlmann, Veit (2010): Reason and Resonance. A History of Modern Aurality, New York.

Feld, Steven (1982): Birds, Weeping, Poetics, and Song, in: Kaluli Expression, Philadelphia.

H.F. (1843): Lowell, in: The Lowell Offering, Lowell.

Kahn, Douglas (1999): Noise, Water, Meat. A History of Sound in the Arts, Cambridge.

Keyes, Sarah (2009): Like a Roaring Lion. The Overland Trail as a Sonic Conquest, in: Journal of American History 96, 19-43.

Lewis, Alonzo (1844): The History of Lynn, Including Nahant, Boston.

M.T. (1842): The Prospect from My Window in the Mill, in: The Lowell Offering, Lowell, 57.

Mumford, Lewis (1961): The City in History. Its Origins, Its Transformations, and Its Prospects, New York.

Parr, Joy (2010): Sensing Changes, Vancouver.

Rath, Richard Cullen (2008): Hearing American History, in: Journal of American History 95, 417-31.

Rosenfeld, Sophia (2011): A Case for Paying Attention to the Historical Ear, in: American Historical Review 116, 316-34.

Schwartz, Hillel (2011): Making Noise. From Babel to the Big Bang & Beyond, New York.

Smith, Mark M. (2001): Listening to Nineteenth-Century America, Chapel Hill.

Smith, Mark M. (2012): The Garden in the Machine. Listening to Early American Industrialization, in: Pinch, Trevor / Bijsterveld, Karin (eds.), The Oxford Handbook of Sound Studies, New York, 39-57.

Smith, Mark M. (forthcoming): When War Makes Sense. A Sensory History of the American Civil War, New York.

Wallace, Anthony F.C. (1978): Rockdale. The Growth of an American village in the Early Industrial Revolution. An Account of the Coming of the Machines, the Making of a New Way of Life in the Mill Hamlets, the Triumph of Evangelical Capitalists over Socialists and Infidels, and the Transformation of the Workers into Christian Soldiers in a Cotton-manufacturing District in Pennsylvania in the Years before and during the Civil War, New York.

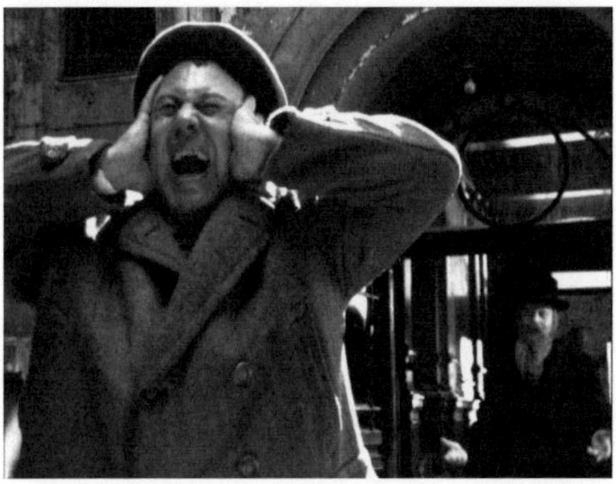

Sounds Familiar

Intermediality and Remediation in the Written, Sonic and Audiovisual Narratives of *Berlin Alexanderplatz*

Andreas Fickers, Jasper Aalbers, Annelies Jacobs and Karin Bijsterveld

1. Introduction

When Franz Biberkopf, the protagonist of Alfred Döblin's novel *Berlin Alexanderplatz* steps out of the prison in Tegel after four years of imprisonment, »the horrible moment« has arrived. Instead of being delighted about his reclaimed liberty, Biberkopf panics and feels frightened: »the pain commences«.[1] He is not afraid of his newly gained freedom itself, however. What he suffers from is the sensation of being exposed to the hectic life and cacophonic noises of the city – his »urban paranoia«.[2]

The tension between the individual and the city, between the inner life of a character and his metropolitan environment is of course a well established topic in the epic literature of the nineteenth century, often dramatized by the purposeful narrative confrontation between city life and its peasant or rural counterpoint.[3] But as a new literary genre, the *Großstadtroman*, or big city novel, only emerges in the early twentieth century, and Döblin's *Berlin Alexanderplatz* is often aligned with Andrei Bely's *Petersburg* (1916), James Joyce's *Ulysses* (1922) or John Dos Passos's *Manhattan Transfer* (1925) as an outstanding example of this new genre.[4] What distinguishes these novels from earlier writings dealing with the metropolis is their experimentation with new forms of narrative composition, often referred to as a »cinematic style« of storytelling. At the same time, however, filmmakers such as David W. Griffiths and Sergei Eisenstein developed

1 Döblin 1961, 13.

2 Cf. Hake 1994.

3 Cf. Williams 1987; Berbig 2011.

4 Cf. Barta 1996.

new forms of filmic storytelling known as »montage«, which were inspired by their reading experiences of realist novels by Charles Dickens or Leo Tolstoy, experimenting with the complex interlacing of parallel narrative plots into a consistent epic format.[5]

This short and admittedly simplified juxtaposition of the montage technique as new filmic language and the emergence of the *Kinostil* as new literary genre brings us to the heart of our motivation for writing this chapter: to reflect upon the ways in which different media, in their narratives and composition, stage the city, and the intermedial relations between these forms of staging. Because of the complexity of the subject, in terms of the range of theoretical and conceptual approaches as well as in terms of the richness of existing scholarship, this chapter concentrates on a single example: *Berlin Alexanderplatz*. Even this self-imposed limitation needs further qualification, for the scholarly work dealing with Döblin and his masterpiece easily fills a library.[6] We will therefore restrict our endeavour to two sets of questions that serve as our basic concern throughout this chapter. First, what does a systematic comparison between the different media forms in which the original novel has been performed (novel, radio play, film and television series) tell us about the different ways in which *the city* has been staged as a symbol of modern conditions of life? And related to this, what are the specific techniques and styles used to create or invent a media-specific dramatization of the story? Second, what is the symbolic role and narrative function of city sounds in these mediated constructions of Berlin as a modern metropolis? How do the sonic representations of Berlin differ in the novel, the radio play and the two filmic adaptations of the subject?

To address these two sets of questions, we will start with a theoretical reflection on the concept of intermediality that informs our comparative approach to analyzing the different media versions of *Berlin Alexanderplatz*: the original novel published in 1929, the radio play *Die Geschichte vom Franz Biberkopf* written by Döblin in early 1930, the film version directed by Phil Jutzi in 1931, and finally the television series directed by Rainer Werner Fassbinder and first broadcasted in 1980. We will describe and analyze the different productions in chronological order. This seems the most appropriate way of structuring the chapter as it allows for a broader contextualization of the different productions and the detection or absence of *familiar sounds* in the different media formats.

While this chapter builds on the excellent studies available on Alfred Döblin, his novel

5 Paech 1997, 122-50.

6 For a detailed bibliography of scholarly work on Alfred Döblin as well as on the different adaptations of the novel in radio, film and television, see Sander 1998. For a detailed portrait of the life and work of Döblin, see the impressive biography by Schoeller 2011.

and its various medial adaptations, such as the impressive study by Peter Jelavich, we also present several new perspectives and more refined interpretations. Our main focus is on the sonic dimension of Döblin's book and its filmic or radiophonic adaptations – a perspective slightly neglected in most scholarship so far. Notably in response to Jelavich's intermedial approach, we aim at offering a stronger comparative perspective, analyzing the various versions of *Berlin Alexanderplatz* not only in their historical contexts but also by looking for intermediality across time. In addition we tone down the somewhat excessive political contextualization and interpretation of the radio play and Jutzi's film by Jelavich, and, at the same time, include into our analysis both the Fassbinder film and a recently realized production of a radio play based on the original script by Döblin. This broadening of the historical and material scope, as our argument will reveal, adds new insights to the rich literature on the topic.

2. Intermediality, Remediation and the Refashioning of Storytelling

That media – in both their form and their content – relate to each other is not a new insight. Yet, the theoretical approach to the various forms of intermedial relationships as a distinct field of research is still relatively young. This is also reflected in the broad, rather unspecific and sometimes contradictory way in which *intermediality* is defined and used as a concept. While scholars in literary studies often use this term in more or less open association with the terms »intertextuality« and »intertext« as developed by Julia Kristeva[7], scholars in media studies often connect it to the concept of »remediation« as introduced by Jay Bolter and Richard Grusin[8] to describe the generic interrelatedness of »old« and »new« media. As Irina Rajewsky has aptly stated, the concept of intermediality has become a *termine ombrellone* – a comprehensive buzzword with many different meanings.[9]

In following the work of Irina Rajewsky, we will use a definition of intermediality that is based on the distinction of three subcategories:

1) Intermediality in the sense of *medial transposition* (for example adaptations of novels in radio or film). This category is a production-oriented, »genetic« conception of intermediality.

7 Cf. Graham 2011.

8 Cf. Bolter / Grusin 1999.

9 Cf. Rajewsky 2002. For a critical discussion of the different meanings of the term see Shail 2010.

2) Intermediality in the sense of *media combination*, which includes phenomena such as opera, theatre, comics, illuminated manuscripts, computer or sound art installations, et cetera. The intermedial quality of this category is defined by the medial constellations that are the result of the very process of combining distinct media.
3) Intermediality in the sense of *intermedial references*. Examples are references in a literary text to a film through the evocation of filmic techniques such as montage, zoom shots or fading. Unlike in *media combination*, in *intermedial references* the referencing medium »thematizes, invokes, or imitates elements or structures of another, conventionally distinct medium through the use of its own media-specific means«.[10]

It is important to note that these categories are not mutually exclusive, and that this subdivision of intermedial practices into medial transposition, media combination and intermedial references is in no way exhaustive. But, as Jürgen Müller has pointed out, this tripartite definition of intermediality is very useful indeed, because it allows us to take into account the hitherto neglected processes of production of media formats while it also invites us to pay attention to the social dimension of media practices.[11] In going beyond the relatively narrow semiotic concepts of intertextual or intramedial references, intermediality in this tripartite definition allows for a broader use of the concept in cultural, social and historical studies. Integration of the material or technological dimension of media products and practices with their social or cultural meanings opens up new perspectives for an intermedial historiography, which is also a central concern of this volume.

In approaching the phenomenon of intermedial transpositions, combinations and references from both synchronic and diachronic perspectives and by focusing on sound, this chapter provides a contribution to the new field of intermedial historiography.[12] This contribution would be rather reductionist, however, if it would neglect the broader historical contexts in which the media products and practices under consideration emerged, developed or changed. When reflecting upon the specific urban environment in which Alfred Döblin worked on his novel *Berlin Alexanderplatz* and on its adaptations for radio and film, one has to be aware of the multimedia reality in which he lived. 1920s Berlin, to follow Janet Ward's account, emerged as European modernism's pre-

10 Rajewsky 2005, 51-53.

11 Müller 2008, 31-46.

12 Cf. Fickers 2012.

mier capital of visual surface spectacles.[13] In this highly mediatized urban environment, all media were involved in struggles for cultural recognition. As Bolter and Grusin have argued, processes of remediation and refashioning of specific media formats are often characterized by rivalry and competition, where technological know-how, industrial strategies and business interests are at stake.[14] Rolf Goebel, in his comparison of the media rivalry between Ruttmann's *Berlin: die Symphonie der Großstadt* and Hessel's *Ein Flaneur in Berlin,* arrives at a similar conclusion:

> Then and now, the interaction between literature and audiovisual media is characterized by a complex dialectic of assimilation and competition, rather than a linear change from the former to the latter, even if the specific terms of this dialectic – styles, themes, ideological perspectives, moral values, etc. – are defined by changes of history, culture, and media technology.[15]

Following Goebel, 1920s Berlin offers a unique synchronic framework for the study of competing cultural media: typewriter, newspaper, telephone, gramophone, radio and film. At a time when Berlin celebrated itself as the hypermodern capital of technological speed, avant-garde arts and rampant consumer fetishism, these media, which originated in different historical contexts, co-existed in constellations of the »simultaneity of the non-simultaneous« (to adapt Ernst Bloch's famous phrase). In many ways, Berlin in the years of the Weimar Republic's cultural innovation can therefore be seen as the

> capital of modernist intermediality, as a metropolis where artists, writers, film directors, and theatre producers experimented with the aesthetic potentials and social effects created by the interaction of various traditional and avant-garde media.[16]

Intermediality, in this sense, is not only a theoretical category for the study of the complex interrelations among different media forms or their intramedial references, but also a lived reality in which new cultural practices emerge. This chapter will in fact address both dimensions by relying on an intermedial historiographical approach.

13 Cf. Ward 2001.

14 Cf. Bolter / Grusin 1999. For an impressive historical analysis of such a process of remediation and intermedial competition between the emerging media of telephone, gramophone, radio and sound film in the 1920s and 1930s, see Wurlitzer 2007.

15 Goebel 2009, 113.

16 Ibid., 111-12.

3. Alfred Döblin, Berlin and the novel *Berlin Alexanderplatz*

When Alfred Döblin, son of a Jewish tailor from Stettin[17], moved to Berlin at the age of ten in 1888, he entered a realm that was – as he declared at various occasions in his later life – to become the »humus« of all his literary and artistic thoughts.[18] This »afterbirth« as he called in an autobiographical retrospective[19] proved to be crucial for his social and intellectual development during the late phase of the Reich. Berlin, he confessed in the same article, »is the fuel that drives my car«.[20] Already in an essay entitled *Modern. Ein Bild aus der Gegenwart* (Modern. A contemporary prospect), written as an eighteen-year-old pupil, Döblin created a flâneur-like portrait of the city as a fluid and vibrant place full of noises and busy people.[21] In 1910, after a successful study of medicine, his first professional experiences as a physician specialized in neurology and several published novels, he joined the editorial board of the newly founded expressionist magazine *Der Sturm* and dedicated his hometown a first allegorical portrait with *Das märkische Ninive* (The Nineveh of Brandenburg). Here, Berlin is described as

> a peculiar city of lust and sin, channelled by railways, a crush of hounded drudges […] only praying to the goddess of time.[22]

With the publication of the short novel *Der Kaplan* (The Chaplain) in June 1914, just before the outbreak of the First World War, Döblin – according to Gabriele Sander – experimented for the first time with a montage-like narrative structure, characterized by a serial listing of disparate impressions and simultaneous perceptions of discrete details

17 Stettin, a trading town in the former German lands of Western Pomerania and today known as the Polish city of Szczecin, is one of the most important port cities of the Baltic Sea located at the Oder River estuary, some 120 km to the North-East of Berlin. Döblin's father had abandoned the family because of another woman and had moved to the United States.

18 In his essay »Berlin and the artists« from 1922, Döblin stated: »all my thinking and intellectual work – whether implicit or explicit – belongs to Berlin […] this ocean of stones that I grew up with is the native soil of all my thoughts«. Quoted in Sander 1998, 75.

19 Döblin, Erster Rückblick, quoted in Sander 1998, 74. Döblin's arrival and »new birth« in Berlin is colourfully depicted in Schoeller 2011 (37-42) too.

20 Quoted in Braunger 2006, 6.

21 Sander 1998, 76.

22 Döblin, *Das märkische Ninive*, quoted in Sander 1998, 77-78.

of reality.[23] The main topic of *Berlin Alexanderplatz*, the confrontation between the subjectivity of the individual and the collective reality of the city is somehow pre-formed in this short novel and was successively developed over the years that followed. After having served as a military doctor at various hospitals in Alsace-Lorraine during the war, Döblin was back Berlin in early 1919 and re-opened a neurological practice in the Frankfurter Allee, just a few blocks away from the Alexanderplatz.

Before Döblin started working on *Berlin Alexanderplatz* in late 1927 he published a number of essays and shorter novels dealing with Berlin as main subject, such as the utopian novel *Berge Meere Giganten* (Mountains Seas Giants, 1924) and the essay *Eine kassenärztliche Sprechstunde* (Consultation hour of a CHI doctor, 1928) where Döblin reflects on his medical practice and the many curious encounters that came with practicing in a working class district.[24] In *Berge Meere Giganten*, set in the twenty-seventh century, the giant city landscape of Berlin is ruled by a tyrant called Marduk. In using the name of one of the Babylonian gods as main character of his fantastic story, Döblin implies an analogy between Berlin and Babylon – the city that carries like no other place the symbolic signifiers of confusion, sin, temptation and destruction. Berlin, which was marked by unprecedented growth since the late nineteenth century and which turned from a residential and bourgeois Prussian city of 900,000 inhabitants in 1871 into the biggest industrial conglomerate of the Weimar Republic with more than 4 million people by 1925[25], served as a modern Babel, a »Chicago on the Spree River«.[26]

Döblin was of course not the first to artistically reflect upon Berlin as a symbol of industrialist modernism. The Futurists in their 1912 Berlin exposition; expressionist painters like Oskar Kokoschka, Ludwig Kirchner and George Grosz; poets like Johannes Becher, Ludwig Rubiner or Georg Heym; dramatists like Bertolt Brecht; or filmmakers such as Fritz Lang and Walter Ruttman – they all struggled to find an appropriate artistic language to come to terms with Berlin as the prototype of the modern condition of men, oscillating between addiction and disgust, inspiration and horror.[27] Yet Döblin was the first to depict and analyze the city using the artistic format of a novel. As Peter Fritzsche

23 Ibid., 79-80.

24 In a lecture that Döblin gave in Zürich 1932 he confessed that in his work as a physician had met a lot of »criminals«, and this had somehow provided him with most of the characters and information for his naturalistic depiction of Berlin and the »Berliners«. See Döblin 1961, 505.

25 For historical statistics of Berlin in terms of inhabitants see Wikipedia (German).

26 Schoeller 2011, 39.

27 See »Nachwort des Herausgebers« in Döblin 1961, 509-510.

has mentioned, this artistic and intellectual examination of Berlin as the modern metropolis has successfully co-produced the modern myth of the city:

> What heightens the fascination is the link between the city as a geographical place and the city as a narrated form. From the beginning, written records and urban existence have gone hand in hand. This correspondence created an imaginary symbolic order that was as important as the city itself.[28]

In *Berlin Alexanderplatz*, Döblin consequently developed a narrative style he had been experimenting with for more than a decade by then. This style, mainly characterized as *Kinostil* or filmic writing in the academic literature on *Berlin Alexanderplatz*, is defined as a narrative strategy that breaks with the »classical« linear construction of the novel with its chronologically or temporarily structured plot and homodiegetic narrator. Instead, the *Kinostil* – as theoretically developed in Döblin's reflections on the aesthetics of the novel in 1913 (the so-called *Berliner Programm*) – was intended as a plea for the re-establishment of the novel as a socially relevant instrument of artistic expression, breaking with the causal identity construction of literary characters in the psychological novels of the nineteenth century. As Eggo Müller and Andrea Melcher have shown, Döblin used film as the most popular symbol of modern aesthetic expression of his times – representing movement, dynamics, acceleration, fragmentation, flow, collectivist production, urbanity, mechanical reproduction and capitalist industry – for critical reflection on the aesthetic form and social responsibilities of literature.[29] Influenced by the observations he made as psychiatrist and inspired by the cinema as the timeliest expression of the edginess and rapidity of urban life, Döblin invented his own style of experimental prose which he described as »Döblinismus«. In a legendary *open letter* to Marinetti, Döblin distanced himself from the linguistic experiments of the Futurists and defended his *Döblinismus* as a narrative prose strategy in order to re-establish the novel as an appropriate form of contemporary literary expression.[30]

28 Fritsche 1996, 4.

29 Melcher 1996, 81-84; Müller 1988, 11-21. As Müller shows, Döblin developed his thoughts in temporal synchronicity with the so-called *Kino-Debatte* (cinema debate) of German intellectuals in the early 1910s. In this debate, prominent figures of the German intellectual milieu like Georg Lukács or Egon Friedell tried to enoble the film as art and as the most authentic aesthetic expression of the modern condition. On the *Kino-Debatte*, see Kaes 1978.

30 Schoeller 2011, 119 and 131-32.

The basic narrative tool for creating a novel which »describes« the urban condition *and* makes it comprehensible and tangible for the reader is the technique of montage. This technique allowed Döblin to experiment with what he called »Tatsachenphantasie«, a factual imagination or poetics based in mimesis. According to Steffan Davies and Ernest Schonfield, Döblin's literary texts emerge from »a hybridisation of discourses: medical, psychiatric, biological, socio-historical, anthropological, ethnological, philosophical, and theological«, masterfully blended in his *Kinostil*.[31]

Following Joachim Paech, the technique of montage in »filmic writing« (*Kinostil*) has at least three forms:

- first, the construction of a meaning based on the combination of disparate narrative elements into a new narrative relationship.
- second, the literary mimesis of a montage-like perception of reality.
- third, the deconstruction of meaningful contexts and their decomposition into disparate elements which – in their heterogeneity – enable new / alternative readings.[32]

If the first two forms of montage can be described as »integrative« techniques of montage, potentially open to the creation of the illusion of a consistent reality by the author, the third form can be characterized as »deconstructive montage«, which aims at reproducing the heterogeneity of our perception of reality at the level of aesthetic production, thereby emphasizing the intrinsic impossibility of constructing a coherent perception of reality.[33] Given the temporal correlation of the emergence of film as a new form of storytelling and the metropolis as exemplary site for the experience of an accelerated modernity, the *Kinostil* proved a most powerful stylistic tool for Döblin, who experimented with montage in several ways.[34]

The most evident form of montage in *Berlin Alexanderplatz* involves the extensive documentary use of text fragments originating from other media, making the novel a primary source for the study of intermedial and intertextual relationships. In collage-like design, Döblin's original manuscript gives evidence of the integration of text fragments from newspaper articles, advertisement boards, poster walls, letters, statistics, popular songs, weather reports, radio announcements, court files, industrial leaflets, administra-

31 Davies / Schonfield 2009, 9.

32 Paech 1988, 129.

33 Ibid., 130.

34 On the temporal correlation and discursive interconnectedness of the city and the film, see Uricchio 1982.

tive forms, window dressings and obituary notices. The integration of these heterogeneous text fragments into the narrative mainly aimed at creating a sense of simultaneity, expressing the disparate and conflicting perception of reality by the main protagonist of the story, Franz Biberkopf. Contemporary critics of the novel already referred to the term of »montage«[35] or used filmic and photographic analogies to describe the new stylistic principle. In a book review for the *Niederdeutsche Heimatblätter*, Chistian Otto Frenzel explained the »absolute verisimilitude« of the novel as a result of the »fast, cinema-like, photographically superposed alternation of images«[36], and Erich Mühsam, a leftist litterateur and political activist, praised the book as »the first and successful example of a written photomontage«:

> No laborious explanations, but a film of short sequences that speak for themselves; life flows by and it does so in front of the cinematographic lens in Franz Biberkopf's brain.[37]

Each of the opening paragraphs of the nine books of which the novel is composed functions as a kind of atmospheric introduction to the next episode, providing the reader with a dense description of the urban environment. The opening passage of the fifth book, in which Biberkopf gets trapped into a thievery and tragically recognizes his failed effort to develop into a »sincere man«, serves as an impressive example of Döblin's *Kinostil*:

> Boom, boom, the steam pile-driver thumps in front of Aschinger's on the Alex. It's one story high, and knocks the rails into the ground as if they were nothing at all. Icy air, February. People walk in overcoats. Whoever has a fur piece wears it, whoever hasn't, doesn't wear it. The women have ice on thing stockings and are freezing, of course, but they look nice. The bums have disappeared with the cold. When it gets warmer, they'll stick their noses out again. In the meantime they nip a double ration of brandy, but don't ask me what it's like, nobody want so swim in it, not even a corps.

35 Walter Benjamin identified the montage as dominant stylistic principle of the novel and praised its epic potential. See Benjamin 1930, »Berlin Alexanderplatz«, quoted in Sander 1998, 146.

36 Ibid., 145.

37 »Da wird nichts umständlich begreiflich gemacht, da rollt ein Film ab in kurzen Bildern, die sich selbst erklären; da flutet das Dasein vorbei, und zwar an der kinematografischen Linse im Hirne des Franz Biberkopf«. See Mühsam 1984, 131-32.

> Boom, boom, the steam pile-driver batters away on the Alex. [...] Everything is covered with planks. The Berolina statue stood in front of Tietz's, one hand outstretched, a regular giantess, now they have dragged her away. Maybe they'll melt her and make medals out of her.
> People hurry over the ground like bees. They hustle and bustle around here day and night, by the hundreds. The street-cars roll past with a screech and a scrunch, yellow ones with trailers, away they go across the planked-over Alexanderplatz, it's dangerous to jump off. The station is laid in our on a broad plan, Einbahnstrasse to Köningsstrasse past Wertheim's. If you want to go east, you have to pass police headquarters and turn downs through Klosterstrasse. The trains rumble form the railroad station towards Jannowitzbrücke, the locomotive puffs out a plume of steam, just now it is standing above the Prälat, Schlossbräu entrance a block further.
> Across the street they are tearing down everything, all the houses along the city railroad, wonder where they got the money from, the city of Berlin is rich, and we pay the taxes.[38]

Like in film, Döblin plays with different »camera perspectives«, offering a montage of Biberkopf's simultaneous perceptions of the dynamic urban reality. Like a cameraman, he zooms in and out, makes tracking and pan shots, and changes the field sizes and lengths of his narrative lens. Obviously, the visual analogies are striking. But it would be short-sighted to reduce the innovative form of storytelling in *Berlin Alexanderplatz* to its analogies of visual montage technologies in filmmaking. The text owes its narrative force and dramatic tension in no lesser degree to the use of a specific language and syntax, characterized by an extensive use of local dialect and onomatopoetic spelling in order to create a naturalistic atmosphere. Both techniques are of crucial importance when it comes to the literary reconstruction of the urban soundscape of Berlin. More than anything else it is Döblin's playful handling of the sonic characteristics of Berlin as a social space, inhabited by people speaking in a very recognizable dialect that make *Berlin Alexanderplatz* a »monument of Berlinism« (*Monument des Berlinischen*), to quote Walter Benjamin. As he put it: »Döblin talks the city. Berlin is his megaphone«.[39] Specifically, according to Toni Bernhart, Döblin relies on three different techniques to make the city sound to its readers.[40] Apart from the abovementioned onomatopoetic

38 Döblin 1997, 216-17.

39 Benjamin 1930, »Krisis des Romans«, quoted in Sander 1998, 144.

40 Cf. Bernhart 2008.

spelling and *O-Ton*[41] literary reproduction of local dialect, the text offers literal descriptions of noises, sounds or music and finally works with sonic presuppositions as a result of the creation of thick sensorial descriptions of specific situations. Examples of the first category are: »Boom, boom, the steam pile-driver batters away on the Alex«[42], quoted above, and

> [...] as Franz closed his eyes, he heard the bells ringing. For several minutes he sat in silence, listened to them ringing: Boom, bim, bum, boom, bim, bam, bum, bum, bim.[43]

Very often, the onomatopoetic spelling is combined with descriptions of sounds or audible phenomena, for example when Biberkopf hears that his girlfriend Mieze was murdered:

> As he thinks of Miezeken something rises in him, fear arises, terror beckons him, there he is, there ist hat mower, Death's his name, hatchet and staff in hand he marches o'ver the land blowing on a little flute, he wrenches his jaws apart, and takes a trumpet, and beats the kettledrum and now it looms, a doom, gloom-black, battering ram, drooms, and softly drooooms.[44]

The use of verbs, adjectives and nouns that explicitly refer to an acoustic event is for sure the most common literary strategy to narrate sound. When Döblin describes the turbulent scenes at the slaughterhouse in the beginning of the fourth book, he creates a dense acoustic impression by using a sequence of nouns or verbs which evoke a dull atmosphere:

> Nothing can be seen, the steam is too thick. But a continuous noise of squealing, snorting, clattering, men's voices calling back and forth, tools being dropped, slamming of lids. [...] Behind there must be the slaughter-pens, there is a sound of smacking, clattering, squealing, screaming, rattling, grunting. [...] Bing, one of them has run in front of his feet, bing, another one. The man is quick, he has given an account of himself, the hatchet has whizzed down, plunged into the lot of them

41 In German *O-Ton* refers to the *original tone* of a live recording on location.

42 Bernhart 2008, 216.

43 Döblin 1997, 267.

44 Ibid., 534.

> with its blunt side, first on the head, then on another. That was a great moment. Kicking, writhing. Flinging from side to side. No longer conscious. Just lying there.[45]

By stressing the fact that nothing can be seen because of the steam, Döblin emphasizes the »blindness« of his protagonist and prepares the reader for his acoustic descriptions of the scene. The sensorial liveliness of these descriptions is further strengthened by the tactile specifications of the butcher's tacit knowledge and the rapid sequence of sonic words (smacking, clattering, squealing) which invoke the trope of the intrusive sound discussed in our introduction. Döblin's snapshot-like literary descriptions of specific situations can evoke an acoustic image in the readers mind, even if he does not refer to a concrete sound event in the narrative. This is what Bernhart describes as »sonic presuppositions«, which pertain to the implicit audibility presupposed by the reader when diving into the narrative.[46] While reading a passage where Döblin describes the hustle and bustle at the Alexanderplatz or a heated debate in a pub from the character's voice perspective, the reader identifies with the character and feels, hears or sees like him. Because of the montage-like construction of the novel and its style of simultaneity, *Berlin Alexanderplatz* has been described as a »sensual event« that

> covers the reader's eye with forms, colours, movements, letters; his ears with exclamations and dull mechanical and human sounds; and the whole body with vibrations and temperatures.[47]

This implicit process of sonification is often made explicit when the author verbalizes his characters' hearing or listening by using verbs that describe these acts. As Toni Bernhart established in a study of the use of such verbs in *Berlin Alexanderplatz*, the novel contains 251 instances of usage of the three main verbs *hören* (hear), *horchen* (hark) and *lauschen* (listen), or compounds or derivations in which they figure.[48] But what is more interesting than this statistical evidence is the question of what is being heard, harked or listened to: What in fact does Franz Biberkopf hear when Döblin makes use of these diegetic references? Surprisingly, perhaps, Biberkopf mainly listens to the voices or speech by his fellow men, rather than to the cacophony of big city life. At least statistically, the great majority of explicit references to acts of hearing, harking and listen-

45 Ibid., 178.

46 Bernhart 2008, 54.

47 Klotz 1987, 390.

48 Bernhart 2008, 55.

ing refers to interactions in which he is either actively or passively involved. Bernhart summarizes his findings with saying that it seems as though the soundscape of the city mainly consists of talk.[49] If somewhat overstated, perhaps, Bernharts findings rightly hint at the importance of the spoken language – and especially the Berlin dialect – as decisive aesthetic and narrative element in the construction of Döblin's *acoustic image* of Berlin.

4. Radio Play. Die Geschichte vom Franz Biberkopf

In early 1930, only a few months after the publication of the novel (October 1929), Döblin started working on a radio play adaptation of his book. Being an enthusiast radio amateur who owned an advanced four-tube superhet receiver by 1928[50], Döblin was also a regular guest on various radio programmes of the Berlin station.[51] He would perform his own works, but he was sought after as a critical mind for talks and discussions about contemporary literature and reflections about the Zeitgeist as well. He debated »cultural Bolshevism« with Friedrich Muckemann, a Jesuit publicist and sharp critic of National Socialism, discussed »The duties of authors in our time« with Bertolt Brecht and other writes, performed improvised storytelling before the microphone with Rudolf Arnheim and Arnold Zweig amongst others, and chatted with his son Wolfgang about the father-son relationship.[52]

Döblin embraced the new technology of radio as an excellent tool for the dissemination of culture among the masses. Against the elitist conception of art and »high culture« defended by many intellectuals of his time, he welcomed the medium of broadcasting for its »multiplier effects« and denied the widely shown cultural pessimism towards the new medium.[53] In contrast to Rudolf Arnheim[54] or Bertolt

49 Ibid., 61.

50 See Jelavich 2006, 76 for a photograph of Döblin tinkering with his radio set. The picture is also used for the cover design of the book.

51 Melcher 1996, 199-201 has documented at least 30 interventions of Döblin on the radio between 1925 and 1932, with a strongest presence in 1930.

52 For a detailed analysis of Döblin's radio presence in Weimar broadcasting, see Melcher 1996, 153-204.

53 Ibid., 143.

54 Cf. Arnheim 1936.

Brecht[55], Döblin never aimed at developing a consistent theory of radio, but he nevertheless had a clear vision concerning the possibilities and limitations of broadcasting as a medium for the dissemination of literature. Although being only third on the list of radio's main or most adequate functions – the first being the dissemination of music, followed by the transmission of news – Döblin asserted that literature had to undergo a fundamental reworking before being suitable for broadcasting. At a conference on *Literature and Radio* in Kassel in the Fall of 1929, jointly organized by the *Reichsrundfunkgesellschaft* (RRG, the federal broadcasting organization of the Reich) and the literary section of the Prussian Academy of Arts, Döblin – as invited keynote speaker – expressed his thoughts on the topic[56]: »Radio is a medium that alters literature. Literature must or would have to change its form in order to conform to radio«. While literature had fallen silent in an unnatural way because of the invention of printing, radio offered a new acoustic medium, able to »revitalize the native soil of literature«. In this context, the »Schriftsteller« (composer of written language) had to become a »Sprachsteller« (composer of spoken language).

The potential of radio, however, to restore the acoustic dimension of literature, appeared to be suitable for a limited number of literary genres only. For Döblin, the characteristics of radio are audibility, brevity, terseness and simplicity, but these characteristics only apply to the lyrical genres and the essay. In contrast, drama and epic, which need »breadth, expansiveness, and flow«, are not suited for radio at all. In Döblin's view, the audible voice does not add anything positive to the fantasy – quite the contrary; the specific sound of a radio voice does in fact restrict the »mental and sensual immersion« into a story. The radio is similarly inappropriate for drama, as a dramatic performance owes its aesthetic quality and social importance to the visual presence of its actors on stage. Theatre, according to Döblin, »is by no means at all broadcasting«; it involves, by definition, »a collective experience«. The theatre productions and drama adaptations that have been aired so far were nothing but »black and white reproductions of a colour picture«. Despite the fact that »radio had the power to address a 100,000 people simultaneously«, he concluded, »it speaks to 100,000 individuals«.

Because of this basic impossibility for broadcasting to assimilate the literary genres of

55 Brecht developed his thoughts on radio in a number of critical essays between 1927 and 1932. Most well-known is his article *Der Rundfunk als Kommunikationsapparat* (1932), where he criticizes the function of radio as a »distribution apparatus« and pleas for a reconfiguration of radio as a »communication apparatus«.

56 A reprint of the proceedings of this conference is to be found in Bredow 1950, 311-66. All quotations in the following passages are short extracts from this lecture as reprinted in Bredow 1950, 311-17.

drama and the epic novel, Döblin ended his lecture with a plea for the development or invention of a truly »folk like radio art«. This new art form, he was convinced, could only be a literary development of the *Hörspiel,* the radio play. Because it blurred the classical boundaries between the different literary genres and it could assimilate music as well as other sounds the radio play offered the potential of a truly new art form:

> I truly believe that only the playful use of lyric, epic and also essayistic elements will enable the production of a real radio play in the future, which also implies the use of the other expressionist means of radio such as music and sound.[57]

With his call for the development of the radio play as a genuine artistic format for radio broadcasting, Döblin echoed similar claims made by avant-garde radio makers such as Hans Flesch[58], Alfred Braun[59] or Walter Bischoff.[60] Many radiophonic experiments of Weimar broadcasting in the late 1920s and early 1930s witness of this search for a new art form. Alfred Braun developed so-called *Hörbilder* (acoustic pictures), piecing together different sounds and noises in a non-narrative and impressionistic fashion in order to characterize a particular setting or idea.[61] Walter Bischoff, in *Hallo! Hier Welle Erdball*

57 »Es ist mir sicher, daß nur auf ganz freie Weise, unter Benutzung lyrischer und epischer Elemente, auch essayistischer, in Zukunft wirkliche Hörspiele möglich werden, die sich zugleich die anderen Möglichkeiten des Rundfunks, Musik und Geräusche, für ihre Zwecke nutzbar machen«. See Döblin 1929, »Literatur und Rundfunk«, in: Bredow 1950, 317.

58 First artistic director of the Südwestdeutsche Rundfunkdienst (Radio Frankfurt), 1924-1929, then head of Berlin based Funk-Stunde (Radio Hour). He was forced from office by the Nazis and partly imprisoned in the concentration camp Oranienburg near Berlin. He vanished without a trace in the outskirts of Berlin in spring 1945. For a detailed discussion of Flesch's innovative impact both on the theory and practice of radio play in Weimar Germany, see Gilfillan 2009, 45-86.

59 Actor and one of the first reporters of Funk-Stunde (Radio Hour) of the Berlin station. There he was head of the literary department (until 1929) and of the news department until he got arrested by the Nazis in 1933. After the war he served as first director of the Sender Freies Berlin (SFB) until 1958.

60 After a career as scenario editor at the Breslauer Städtische Bühnen, Bischoff was artistic-director and director general of the SchlesischeFunkstunde (Silesian Radio Hour, 1925-1933). He too had to resign from all functions after the Nazis came to power and was arrested. After the war, he became the first director general of the Südwestfunk in Baden-Baden. For biographical notes of all three personalities, see Bräutigam 2005.

61 See Jelavich 2006, 80.

(Global Station Calling, 1928), experimented with an additional principle of composition, based on stringing together a sequence of disparate scenes, such as a telegraph office, a transatlantic steamer, an African village, and so on.[62]

A masterful realization of Braun's concept of *Hörbilder* is of course Walter Ruttmann's acoustic montage *Weekend* (1930). After the success of his avant-garde silent film *Berlin, Symphony einer Großstadt* (1927), Ruttmann created an acoustic portrait of a weekend in Berlin, adapting his film editing techniques to sound broadcasting. This eleven-minute experimental »acoustical film«, as Alfred Braun called it, consists of 240 sound bites of less than three seconds on average and explored the interface of film and broadcasting technologies through the use of montage, sound sampling techniques and editing »to capture and rearrange the fleeting and oftentimes unnoticed sounds of daily experience«.[63] The intellectual debates about the radio play as genuine art form coincided with the first experiments of utilizing the new technology of optical sound recording on film stock for broadcasting. This new emerging genre of the *Hörfilm* was not only a radical innovation in terms of its aesthetic and narrative characteristics; it was also a technical innovation and practical challenge. Both Bischoff and Ruttmann used the *Tri-Ergon* process which had been in use by experimental filmmakers since the mid-twenties. The fixation of sounds on film tape enabled for the first time the editing of sound bites in a montage-like manner, whereas earlier radio play experiments with pre-recorded sounds on discs had struggled with the »unnatural« combination of live and recorded sound. It was Hans Flesch, the avant-garde pioneer of Weimar radio and director of the Frankfurt station, who recognized the radical innovative potential of such technologies for the medium of broadcasting and for the radio play in particular. During a meeting of the programme committee of the *Südwestdeutscher Rundfunkdienst* in Wiesbaden in early June 1928, Flesch held a flamboyant plea for the use of recording technologies in the production of radio programmes. As a »mechanical instrument« (!), radio could only find its characteristic function by fully complying with the mechanical nature of the medium. The radio play in particular requires absolute precision, which could only be achieved through sound film:

> By recording a radio play on sound film, the director can – by means of cutting, cross-fading, and splicing – create a piece of art to be presented to the listener in the evening after he is fully pleased with it.[64]

62 Ibid., 81.

63 Gilfillan 2009, 3.

64 Flesch 1928, 31-63. Quoted in Stoffels 1997, 719.

In a similar way, Bischoff noted that »acoustic dramaturgy is unthinkable without technical dramaturgy«.[65]

While many broadcast professionals reacted with scepticism to Flesch's programmatic statements or Bischoff's experimental broadcasts, their vision perfectly resonated with Döblin's ideas about what an original radio play should be. Döblin himself had made first contacts with the genre of radio play in 1929, when his drama *Lusitania* (1919) had been reworked as a radio play by Hans Peter Schmiedel and broadcasted – opening with an announcement by Döblin himself – under the new title of *Der Untergang der Patagonia* (The sinking of the Patagonia) by the Leipzig station in October 1919.[66] After the big success of his novel, Döblin – most likely by invitation of the new director of the Funk Stunde AG, Hans Flesch – started working on the script for a radiophonic adaptation of his novel. Already in May 1930, Flesch announced the production of the play which can be interpreted as a symbol of the great prestige that he accorded to the project. Another indication of the Radio Hour's high regard for Döblin's work was, according to Peter Jelavich, the fact that they hired the well known actor Heinrich George to lend his voice to the character of Franz Biberkopf.[67] George, who was to dominate the screen version of *Berlin Alexanderplatz* a year later, was also meant to become the omnipresent voice of the radio play.

But how to rework a novel of 529 pages into a radio play of approximately 80 minutes? A massive abbreviation of the story was needed of course, but to remediate the complex novel into an »acoustic picture« or »sonic photomontage« asked for more than a drastic shortening of the narrative. Erich Mühsam, who had praised the novel as the first successful literary »photomontage«, argued in an essay entitled *Franz Biberkopf and Radio* that the radio play should simply replicate the acoustic montage of the novel. Just as Döblin's manner of writing was »a fabulously subtle recording apparatus« that was able to create a lively impression of Biberkopf's feelings and of Berlin's turbulent life,

> the radio microphone only needs to listen in wherever a scene from *Berlin Alexanderplatz* happens to be playing, without any outside direction.[68]

65 Quoted in Jelavich 2006, 87.

66 Döblin decided to change the title of the play because he feared anti-semitic outbursts similar to the ones he had faced when *Lusitania* was premiered in Darmstadt in 1926. The title of the drama referred to the British passenger ship that had been torpedoed by a German submarine in 1915. The fact that the sinking of the Lusitania contributed to American intervention in the war made it a sensitive topic even after the war.

67 Jelavich 2006, 98-99.

68 Mühsam 1930, quoted in Jelavich 2006, 100-01.

In short, the play needed to radiate the novel's spirit and language, by being *biberköpfisch* (Biberkopf-like).

According to Peter Jelavich, the necessity to produce a play in line with the characteristics that Döblin himself had formulated in his Kassel lecture in 1929 – audibility, brevity, terseness, and simplicity – confronted the author by one choice in particular: he needed to decide whether the play should focus on Berlin or Biberkopf as protagonist.[69] Döblin chose the latter – a choice which had a huge impact on the play's narrative composition and dramatic staging. Although he provides us with no traceable evidence for his motivation to foreground the story of Biberkopf, the play's revised title (*Die Geschichte vom Franz Biberkopf*; The Story of Franz Biberkopf) and its narrative structure clearly reveal the secondary role of Berlin. What are the implications for the radio play's representation of the urban soundscape of Berlin? How are urban sound events used to dramatize the narrative?

Before addressing these questions, it is important to point at the fact that the play was never aired during the years of the Weimar Republic. The first broadcast of the play, based on the original script of Döblin, only happened in 1963, six years after his death.[70] Curiously enough, one had to wait until the fiftieth anniversary of Döblin's death in 2007 before the play was realized in a version that fully concords with the original radio manuscript written by Döblin in 1930.[71] To this day, there has been no consistent account explaining the cancellation of the play's broadcast, which in advance was prominently advertised in radio journals and on the radio. While some refer to inopportune political circumstances that confounded the airing of the play because of its leftist undertone, others refer to statements by Flesch and Döblin, who publicly claimed that the play wasn't ready for dissemination yet. Schwitzke and Jelavich strongly push for a political interpretation of the cancellation referring to the sweeping success of the National Socialist Party in the September elections of 1930. The NSDAP reached 18.2 %

69 Ibid., 100.

70 A first transmission was realized by the Berliner Rundfunk of the GDR on 11 August 1963, to be followed by a transmission on the Norddeutschen Rundfunk on 15 September the same year. See Sander (1998), 225. Both versions were based on the original manuscript of the play written by Döblin and on the transcription of the pre-recorded prototype-version of the original play by Wolfgang Weyrauch. For the transcript of the recorded version, see Prangel 1975, 199-236, for the edited version of the manuscript by Döblin, see Schwitzke 1976.

71 The play was a co-production of the Südwestrundfunk (SWR), the BayrischerRundfunk (BR) and the Rundfunk Berlin Brandenburg (RBB) and aired in 2007 on 24 June (BR 2 and SWR 2) and 26 June (RBB). The complete play is available on CD as a bonus to the film on the DVD-edition sold by Arthaus.

of the votes, 15.7 % more than in the elections of two years earlier, and became the second strongest party in the Reichstag.[72] Earlier anti-Semitic attacks against Döblin and deletion of politically critical words or passages (for example the opening dialogue between Job and »a voice« and the slaughterhouse scene) are interpreted to be the result of ideologically motivated censorship and a sign of the rising influence of right-wing extremists in Weimar broadcasting. To counter this political interpretation of the radio play's cancellation, Sander and Melcher argue that Döblin himself announced it on radio the day it was scheduled for transmission (30 September 1930) and that this suggests he was motivated by artistic reservations, rather than political ones.[73] Döblin said that the play in its actual state was not »mikrophongeeignet« (ready for the microphone) and that whole passages needed to be completely reworked. He explained they had started experimenting with pre-recording some passages on sound film and discs – an experience which had opened up completely new possibilities in terms of sound design and narrative direction. That a complete first version of the play was pre-recorded indeed and survived in the German *Rundfunkarchiv* strongly indicates in our view that the cancellation happened due to artistic considerations.[74]

The 2007 production gives us a vivid impression of how Döblin might have wanted his play to sound. It opens with a dialogue between Job and »a voice«, staged as a prologue recorded on a shellac disc. The voice whispers Job's name three times, and Job inquires »who is calling? [...] Who is it? ...Who is asking«? »I'm only a voice«, to which Job replies, »A voice comes from a throat«. The voice continues asking what he fears most and why he doesn't accept any help. Job replies that nobody wants to help him – neither god nor the devil. »And you«, the voice asks, »do you help yourself«? »I can't«, Job replies. »You must«, the voice answers laughingly before it fades away [...] Because of its »radical indeterminacy« the opening offers a brilliant way to begin the story,

> inasmuch as every radio listener likewise hears a voice that does not emanate »from a throat«. [...] Thus at the very outset of the script, listeners are warned to be suspicious of whatever they hear on the airwaves.[75]

More so, the dialogue between Job and »the voice« (which reappears twice during the play) – although it is not an accurate reflection of the biblical figure of Job – mirrors the

72 Schwitzke 1976; Jelavich 2006.

73 Sander 1998; Melcher 1996.

74 See the copy at the German Broadcasting Archive, Frankfurt, signature C 457.

75 Jelavich 2006, 104.

fragile mental and emotional state of Biberkopf and underscores his hubris.[76] The playful association between the figures of Job and Biberkopf is reinforced by the fact that both parts are spoken by the same actor.[77] Right from the beginning of the play, the listener is confronted with a complex narrative construction, characterized by a constant shift between author and character voice, thereby blurring the boundaries between the diegetic and non-diegetic narrative perspective.

But what about the sonic representation of the city? Does the play offer the »acoustic images« of Berlin that Döblin was hoping for, despite his decision to have Biberkopf as the story's protagonist? Indeed it does. By constructing various »acoustic bridges«[78] among sounds and voices of different realms, the play offers a fascinating radiophonic portrait of Berlin as *locus delicti* of the story. Already the opening scene right after the prologue puts the listener in the position of a pedestrian walking on the streets of Berlin. He encounters voices from different spheres. A cacophony of vendors (from which Biberkopf's voice stands out) and newspaper men draw the listener immediately into the heart of the city. The whole scene is acoustically underpinned with music and *O-Ton*-street noises, creating an interesting tension between a naturalistic impression and a deliberate deconstruction of the diegetic perspective. This principle of mixing pre-recorded background noises (for example of a pub) with music and using them simultaneously as acoustic background for the play's monologues and dialogues make the 2007 version an intriguing example of a playful and complex narrative, constantly blurring the boundaries between realistic reportage and fantastic narration, between reality and psyche.

The strongest signifier with regard to the play's topographical setting is the extensive use of Berlin dialect, however. The main voices of the 2007 play are all taken from well known Berlin actors, such as Andreas Leupold (Biberkopf / Job), Andreas Schmidt (Reinhold) and Jule Böwe (Mieze). Their Berlinesque voices have probably the strongest impact on making the play sound »authentic« in terms of its spatial setting. But in order to maintain the ambiguity of character identity – the eliding and confusion of voices were

76 For an analysis of the meaning of Döblin's references to Jewish religion and mythical figures in *Berlin Alexanderplatz*, see Bayerdörfer 1985 and Bird 2009.

77 In the 2007 version the role of Biberkopf (and Job) is spoken by Andreas Leupold.

78 The concept of »acoustic bridge« stems from Rudolf Arnheim's theory of radio (1936). Arnheim argues that the lack of visual clues in radio allows the creation of acoustic bridges that arise between all kinds of different sounds: voices, recitations, discussions, songs, music, noises – »what hitherto could exist only separately now fits organically together: the human being in the corporeal world talks with disembodied spirits, music meets speech on equal terms«. See Arnheim 1936, 199.

a fundamental principle of composition in the novel – Döblin, in the view of Jelavich, resorted to »an obfuscation of sociolinguistic boundaries«:

> To rescue the polyvocality of his novel, Döblin had to emlay, quite literally, multiple voices; but whereas the printed text allowed a blurring of identities, the obvious distinctions among the voices of actual actors and actresses forced the roles to be differentiated. [...] In particular, when the voices spoke *Hochdeutsch* they maintained their distance from Biberkopf; but the more they slipped into word games, rhymes, and Berlin dialect, the more they assimilated themselves to his thoughts and character. While the voices of the actors and actresses could not help but remain distinct, they could meld into each other across the »acoustic bridge« by varying their manner of speaking, by becoming *biberköpfisch*.[79]

This »sociolinguistic slippage« among voices, as Jelavich points out, was paralleled by the use of music to bridge the realms of the play. As in the novel, the radio play – both in its original and in its 2007 version – evoke many popular tunes of the day. Tunes like *Seit wann bläst deine Großmama Posaune* or *Ich hab' mein Herz in Heidelberg verloren* feature prominently. When Biberkopf first meets Sonja (the prostitute who later becomes his girlfriend Mieze), they start singing a popular tune from the operetta *Eine Frau von Welt* in duet:

> Schwör mir keine Treue, leist mir keinen Eid [...] . denn es reizt das Neue jeden mit der Zeit. Heiße Herzen geben niemals Ruh, suchen frischen Antrieb immer zu. Schwör mir keine Treue, weil ich mich zerstreue – gerade so wie du.[80]

Popular music as a major form of urban entertainment and leisure culture functions as an important intermedial element both in the novel and in the radio play, but its impact on the imagination of the »reader« is even stronger in the acoustic medium of the radio (or sound film).

Döblin's most innovative contribution when thinking about how to extend the artistic possibilities of radio in his *Story of Franz Biberkopf* however is his idea to give a voice to mute objects. Probably the most startling scene of the whole play is the housebreaking passage where Biberkopf has to keep cave. The scene is introduced by the voices of the two cars – one *Fiat* and one *Opel* – which will have to drive the Pums gang to the storehouse:

79 Jelavich 2006, 111.

80 The song was originally performed by Max Hansen and Edith Schollwer (1927).

What kind of car are you?
An old Opel, and you?
A Fiat! I've been standing and waiting here for 3 hours already…
What are you waiting for, Opel?
It is not hard to imagine…
Kaiser Wilhelm Street: they're out to play a trick on us again!
Sounds familiar!
Watch out! They are coming.[81]

After this opening dialogue between the two cars, the members of Pums gang get into the cars whose engines are roaring. The whole scene – the drive to Kaiser Wilhelm Street, the housebreaking, Biberkopf's realization that he was taken for a ride, his rising antagonism that grows into a hysterical laughing when they try to escape, and finally his brutal being thrown from the car and being overrun – is constantly commented upon and interpreted by the voices of the two cars. Their duet-like dialogue – with the alternation of the two voices reinforced by the music's rhythm – serves as a dynamic motif, as well as a powerful narrative element in the scene's dramatization. Both the Opel and the Fiat express their disapproval of the criminal activities and show a great deal of sympathy for Biberkopf, especially when he is thrown from the car. The scene ends with the fading sound of the two cars running away from the overrun body of Biberkopf – who, in the next sequence – wakes up in the hospital and learns that the doctors had to amputate his right arm in an emergency operation [...]
Although Döblin during his lifetime has never had the chance to hear a broadcast of his radio play, the historically accurate 2007 version of the play tellingly witnesses of the ageless modernity and narrative power of his original manuscript and of his successful ambition to contribute to the development of the radio play as a genuine acoustic art form.

81 Was bist denn Du für'n Wagen?/'N alter Opel, und Du?/Ein Fiat! Ich steh' hier schon seit drei Stunden und warte…/Weeste, worauf Du wartest, Opel?/Kann mir det schon denken…/Kaiser-Wilhelm-Straße: die wolln' neuet Ding drehen!/Is mir nischt Neuet./Pass uff – sie kommen/Transcript of the dialogue (2007 version of the play) by Andreas Fickers. In the original recording of 1930, one of the car voices was spoken by the brother of Alfred Döblin, Hugo Döblin.

5. Audiovisual Remediation. *Berlin Alexanderplatz* by Phil Jutzi (1931)

Already during his work on the script for the radio play in early 1930, Döblin was approached by Emil Jannings – arguably one of the most prominent actors of the German film scene – in order to talk about the possibility of a film adaptation of *Berlin Alexanderplatz*. Many critics of the novel underlined its *cinematic* quality, and the montage structure and *Kinostil* of Döblin's text seemed predestined for a filmic adaptation of the story. But while Döblin had been an enthusiastic radio amateur and devoted radio listener, he was critical of film as a mass medium. Although film had fascinated him from the beginning as a new artistic form of visual storytelling, he was disgusted by its commercialization and virtually exclusive exploitation as mass entertainment industry.[82] For Döblin, cinema was first and foremost a social institution that needed to fulfil a social function. But, as he conceded in an interview on the occasion of the production of *Berlin Alexanderplatz* in August 1930, »only a changed author can change the film«.[83] In Döblin's view, a film had to »engage with social things« and the directors and script writers in return should not »fear reality« and learn to communicate their social messages to the mass public., next to each other]. It was Jannings who was instrumental in mollifying the cinematic suspicions of Döblin[84], but another factor was of greater importance in encouraging Döblin to consider filming the novel: the emergence of *talkies* in the

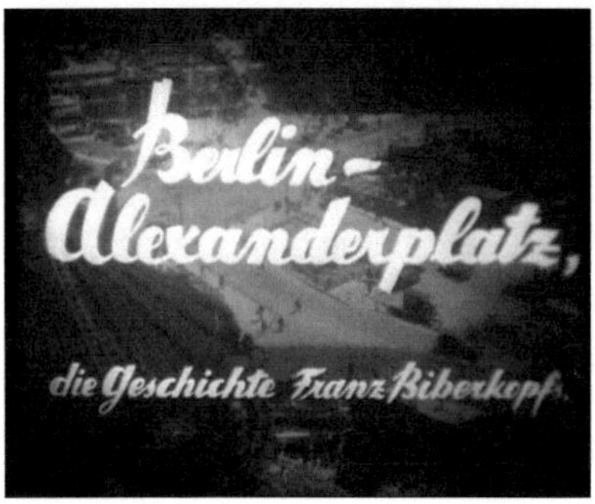

Figure 4: Still of the film *Berlin Alexanderplatz* (1931) by Phil Jutzi: Title strip.

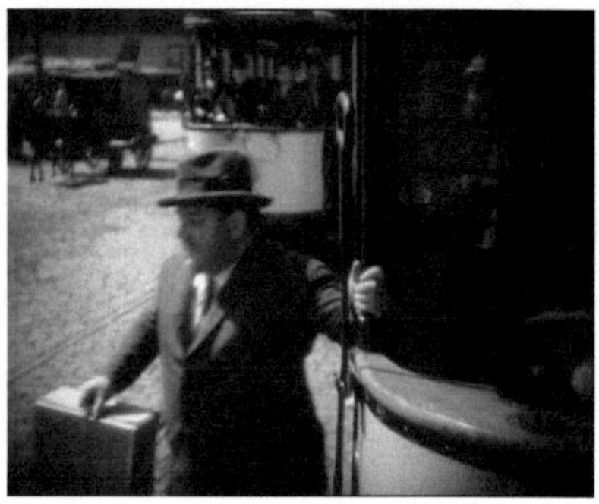

Figure 5: Still of the film *Berlin Alexanderplatz* (1931) by Phil Jutzi: Biberkopf jumping off the tram shortly after being discharged from prison.

82 See Jelavich 2006, 194, and Melcher 1996, 25-50.

83 Döblin 1930, cited in Sander 1998, 228.

84 Jelavich 2006, 197.

late 1920s. Since the mid-1920s, UFA already possessed the patent to the *Tri-Ergon sound film technology* developed by Hans Vogt, Joseph Engl and Joseph Masolle. But a 1925 short experimental film, *Das Mädchen mit den Schwefelhölzern*, was so disappointing that one year later the company decided to sell the patents to a Swiss conglomerate.[85] It was only after the American success of *The Jazz Singer* (Crosland), in 1927, that the sound film seemed a commercially and artistically realizable product. In 1928 two German companies immediately started working on a technology that could rival that of the American combination ERPI / Warner Brothers. Tobis (Tonbild Syndicate A.G.) acquired the Tri-Ergon patents and several hundreds more, while *Klangfilm* was a collaboration between the electrical manufacturers Siemens and Halske A.G. and AEG (Allgemeine Elektrizitäts Gesellschaft). Rather than competing between themselves the two companies decided to merge to be more competitive with ERPI / Warner Brothers.[86] On March 12, 1929 the first German sound film was released: *Melodie der Welt.* This film by Ruttman is a musical journey around the world in which Berlin does not play a part, as in fact is true of his later sound films. The first visits to Berlin in the history of the sound film date from 1931: *M – Eine Stadt sucht einen Mörder* (Lang), *Emil und die Detektive* (Lamprecht) and *Berlin Alexanderplatz* (Jutzi).

The combination of the reality effect of moving images with the authenticity of spoken language and the dramatic potential of music seemed to offer the right technical and aesthetic possibilities to realize a filmic portrait of the doom of Biberkopf, who – just as in the radio play – should become the story's protagonist.[87] After Arnold Preßler from Allianz-Tonfilm GmbH purchased the film rights for *Berlin Alexanderplatz* and hired Heinrich George to play the Biberkopf role, Döblin started writing the screenplay together with Hans Wilhelm, a professional script writer who at that point had mainly composed light comedies and entertainment films.[88] Preßler engaged Phil Jutzi to direct the film. Jutzi began his career filming »low budget Sauerkraut Westerns« in the late 1910s, but he subsequently developed into a highly recognized exponent of the social realist school of leftist filmmakers.[89] As a result, *Berlin Alexanderplatz* was going to be

85 Hake 2008, 55.

86 In March 1929 the merger was finalized. See Gomery 1976, 51-54.

87 Müller 1992, 95-96.

88 Sander 1998, 229.

89 Jutzi adapted Eisenstein's *Battleship Potemkin* for German distribution (1926) and joined the Communist Party two years later. With films like *Um's tägliche Brot: Hunger in Waldenburg* (1928) and *Mutter Krausens Fahrt ins Glück* (1929), Jutzi invented a Marxist version of docudrama. According to Jelavich, »a powerful, moving, and revolutionary work«. See Jelavich 2006, 201-08.

a realistic film. But this realistic turn obviated the symbolism and spiritual multidimensionality that pervaded the novel – much to the deception of most film critics at the time.[90] Because most of the film was shot at the UFA studies in Neubabelsberg while the main actors and actresses had little or no experience with sound film, the final result was a rather stereotypical visualization and sonification of Berlin. Moreover, the film's main character was played by Heinrich George, whose stage persona, according to Jelavich, »had become a stereotype«.[91] The fact that *Berlin Alexanderplatz* was Jutzi's first talkie may also have contributed to the rather conventional end product, which Siegfried Kracauer bitterly labelled »an average gangster flick«.[92]

But what about the sonic representation of Berlin in Jutzi's film? Against the overwhelmingly critical interpretations of the film in academic discourse, Eggo Müller shows that the film offers a number of innovative narrative elements, especially when it comes to the use of sound. As his detailed analysis of the screenplay demonstrates, Döblin and Wilhelm thought very carefully about how to use the new technology of sound-on-film recording as a stylistic tool and narrative element in its own right.[93] This already becomes evident in the opening scene of the film showing Biberkopf's journey into the city, emphasizing the »danger« of the metropolis and Biberkopf's mental fragility. When he gets on the tram to the city, the music – as discussed in this volume in our »Shifting Sounds« chapter – is first punctuated by and then merges with the sounds of traffic: the tram engine, the tram bells and car horns. At the climax of the scene the sounds and sights of the city become too much for Franz, and he jumps off the tram. In a remarkable contrast to the score of *Berlin, die Sinfonie der Großstadt*, the soundtrack of *Berlin Alexanderplatz* offers a telling example, as Jasper Aalbers has claimed, of the »noisification of music«.[94]

This shift from music to noises is adhered to consistently in *Berlin Alexanderplatz*. While there is plenty of diegetic music in scenes taking place in cafes and night clubs and during a wedding, it is not part of the big outdoor scenes in the film. During the scenes at the titular Alexanderplatz, when Franz is working as a street vendor, the film presents us with the soundscape of a modern city. Trains and trams arrive at and depart from the big train station; buses and cars drive by and sound their horns. Behind Franz there is

90 For a detailed description of the contemporary reactions that the film provoked, see Sander 1998, 234-44.

91 Jelavich 2006, 222.

92 Ibid., 223.

93 Müller 1992, 103-05.

94 Aalbers (forthcoming).

construction work taking place, filling the square with its noises. Only from far away the soothing sound of church bells is heard to announce lunchtime. Biberkopf has to raise his voice to be heard by the crowd surrounding him, and he urges people to come closer, so that he may still have a voice left the next day. All in all, the noisification of music in the opening sequence and the conscious construction of the soundscape of Berlin in the rest of the film lead us to suggest that, in direct contrast to Walter Ruttman's *city symphony*, Phil Jutzi's *Berlin Alexanderplatz* certainly presents a cacophony of big city life.[95]

6. The Serialization.
Rainer Werner Fassbinder's *Berlin Alexanderplatz* as TV series (1980)

Viewers had to wait another forty-nine years before they could watch a second attempt at visually capturing Döblin's novel. This new version of *Berlin Alexanderplatz*, realized by the German avant-garde filmmaker Rainer Werner Fassbinder, confronts us with yet another form of adaptation: the translation of the novel into a television series. Fassbinder, now referred to as one of the outstanding film directors of the New German Cinema of the 1970s, had directed several films based on epic or dramatic artworks when he decided to embark on the *Berlin Alexanderplatz* project. It turned out to be the monumental late work of Fassbinder, consisting of a series of thirteen episodes and an epilogue with a total length of fifteen and a half hours of film.[96] Commissioned by the Westdeutscher Rundfunk (WDR) and co-produced by the Bavaria Film Studios in Munich and the Italian public service broadcaster RAI, the film was the most expensive television production ever realized on German television so far. With a total budget of 13 million marks, the film was shot in 154 days between June 1979 and April 1980 and first broadcasted between October and December 1980 on WDR.[97]

95 See Aalbers (forthcoming).

96 For a detailed description of the different episodes and information on the production process and actors involved, see the 60-page booklet »Fassbinder: Berlin Alexanderplatz remastered«, which goes with the Arthaus DVD edition (2006) of the film.

97 Lorenz, 2006, 21-23. Parts of the film were shot at original settings in Berlin (mainly scenes playing in the backyards or staircases of the typical Berlin tenement houses or at underground stations), but all street settings were shot at the Bavaria studies in Munich were the »Berliner street«, Biberkopfs apartment and his local had been rebuilt. See Braunger 2006, 33.

Realizing this film had been an old dream of Fassbinder, who declared in an autobiographical essay published a few days before finishing its shooting in the German weekly *Die Zeit* that his reading of the novel at a younger age has had a significant effect on the further course of his life.[98] This almost existential importance of the work for Fassbinder indicates that the filmic adaptation expectedly showed a highly subjective reading of the novel. Fassbinder never intended to produce a film that would stay close to the original, but he decidedly opted for a biographically motivated interpretation of the text, using the main protagonist Franz Biberkopf as an agent to stage the complex and obscure condition of his own haunting biography.[99] As is true of most of his other work since the late 1960s, Fassbinder's *Berlin Alexanderplatz* provoked heated debates in both the popular and quality press and developed into *the* media event of 1979 in the Federal Republic of Germany.[100] While most of the professional critics praised the film as a milestone of avant-garde filmmaking – the film also being a big success at the film festivals of Venice and Oberhausen in September 1980 – the popular press (Springer press) sharply attacked the film as a product of violence, sex, and morbid fantasies of the director.

The response in the quality press is more relevant to the subject of this chapter. Critics specifically discussed the intermedial relationships between the film and the novel and they focused on the aesthetic and narrative strategies employed by Fassbinder to dramatize the many internalized conflicts, fears and perceptions of the film's protago-

98 »[…] mein Leben, gewiß nicht im Ganzen, aber doch in einigem, manchem, vielleicht Entscheidenderem als ich es heute zu überblicken vermag, wäre anders verlaufen, als es mit Döblins Berlin Alexanderplatz im Kopf, im Fleisch, im Körper als Ganzes und in der Seele, lächeln Sie meinetwegen, verlaufen ist.« See Fassbinder 2006, 5.

99 Sander 1998, 245-56. The homepage of the Fassbinder Foundation offers a detailed overview of his life and works, and documents the rich intellectual debates and discussions that his films, theatre plays and public statements evoked. »His flamboyant and at the same time seedy life-style, his openly displayed and well advertised homosex- uality, and at the same time life and love to women [sic], the scandals, public outrages and bouts of self-pity ensured that in Germany itself Fassbinder was permanently in the news, making calculatedly provocative remarks in interviews, which nonetheless were usually shrewd and to the point. His work often received mixed notices from the national critics, many of whom only began to take Fassbinder seriously after the foreign press had hailed him as a genius.« See Elsaesser, The Fassbinder Foundation, 2010.

100 For a detailed list of the many reviews and critical articles both in press that preceded, accompanied and followed the broadcasting of the series at the public service television station of WDR, see Biesenbach 2007, 656-59.

nist, Franz Biberkopf, so brilliantly performed by Günter Lamprecht. While this concentration on the inner life of Biberkopf has been criticized as a re-psychologization of the main topics of the novel and, for that reason, as a reduction of its portrayal of societal complexity and urban conditions to a mere portrayal of the central character's psychology,[101] others interpret the radical appropriation and personal interpretation of Döblin's text by Fassbinder as a perfect example of a post-modern aesthetic, demonstrating the subjectivist and constructionist nature of all artistic expression.[102] But in contrast to the radio play and the Jutzi film of 1931, Fassbinder's focus on the character of Biberkopf was not chosen because of narrative simplification and genre-specific conventions of storytelling. Instead, Fassbinder consciously broke with established conventions both concerning the length of a classical feature film and concerning aesthetic and moral norms in the public service television culture of the Federal Republic.

The most radical difference between Jutzi's filmic adaptation of Döblin's novel and Fassbinder's approach is certainly their length. While Jutzi fitted the story into a standard feature film of his days with a duration of 84 minutes, Fassbinder produced a monumental film of nearly 900 minutes, expanding the narrative time span of his story into a series of 13 episodes.[103] This extraordinary attempt to break with the conventions of filmic storytelling is, according to Susan Sontag, the radical innovation in Fassbinder's approach and the main condition of its success. In comparing Fassbinder's project with Erich von Stroheim's failed ambition to produce a complete filmic adaptation of Frank Norris novel *McTeague* in 1924[104], Sontag praises Fassbinder's film as the first »real« filmic adaptation of a great big city novel ever.[105] The pure length of the film is of course no sufficient condition for making the film a great movie, but certainly a necessary one. The serialization of the film into an episodic narrative constitutes – in the eyes of many critics – an artificial fragmentation of the storyline and destroys much of the immersive power of the film. According to Fritz Göttler, Fassbinder dreamt of cinema that sucked the viewer into the fictional reality of the story, of films that »tears holes in the life of

101 See for example Burger (1987), in: Sander 1998, 250-51.

102 See Haag, »Fassbinder verfilmt Berlin Alexanderplatz«, in: Sander 1998, 254-55.

103 Episode number 14, the »epilogue«, has no equivalent in the novel. It is a highly allegorical and provocative filmic commentary on the main topics of the novel by Fassbinder.

104 When the silent film called Greed was first shown in theatres, it had a total length of 2 hours and 20 minutes. Stroheim originally produced a version with a total length of nearly ten hours, the film was radically cut down by the producers (Metro-Goldwyn-Mayer) in order to fit it into the conventions of feature films of the time (because of financial reasons). See imDb.

105 Sontag 2007, 48-53.

their viewers«. In that sense, Göttler interprets Berlin Alexanderplatz as a »masterpiece of filmic somnambulism«.[106] And indeed, where Fassbinder's opus magnum was shown in movie theatres in full length (mainly in a two-day or three-day screening as in New York during several weeks in 1983), the reactions of the public were much more positive than in Germany, where the televised episodes where broadcasted in a late night timeslot (starting at 9:30 pm) and at weekly intervals.[107]

One could argue, however, that the »forced« format of serialized production (WDR commissioned the film as a television series) also allowed for a stronger association of the filmic adaptation with the nine books of the original novel.[108] Fassbinder even strengthened this association by inserting lengthy quotations from the novel, read by Fassbinder and presented in an off-screen voice mode. The titles of the 13 episodes, all based on passages from the novel, were prominently visualized in old German letter type in the front credits of each episode, and recurring motifs like the abattoir scene or the appearance of »Schnitter Tod« (Grim Reaper) are accompanied by chroma keying of text passages from the novel. But the Fassbinder production shows many other intermedial references such as the incorporation of original photographs (of Berlin or the abattoir), newspapers articles and advertising columns, graffiti on frontages, billboards and board partitions, and so on. The most prominent intermedial references Fassbinder plays with, however, are of a musical or acoustical nature. While Biberkopf's close relation with music is staged in many singing scenes, the sound design of the production is characterized by a strong presence of background noises, often dominated by music. Music is used in both a diegetic and non-diegetic way. Biberkopf very often sings or hums to popular tunes from the gramophone or radio in his home. But he also sings in public – for example in episode two (»Wie soll man leben, wenn man nicht sterben will« / How to live when you don't want to die?), when he engaged in a singing contest with a group of communists in his local. Challenged by their singing of the *Internationale*, he responds with *Die Wacht am Rhein* – a German folk song written by 1840 von Max Schneckenburger and set to music in 1854 by Carl Wilhelm. During the Kaiserreich (1871-1918), the song was extremely popular and acquainted – albeit never official – nearly the status of a national hymn. Biberkopf already performed the song in the first episode when, in a mood of panic and disorientation after his disimprisonment, he fled into a courtyard of a tenement. In singing the song with a strong voice, he regains self-confidence and reveals his relief.

In addition to the diegetic performances of the singing Biberkopf, non-diegetic musical

106 Göttlich 2007, 32.

107 Lorenz 2006, 23.

108 The different chapters of the original novel were entitled »books« by Döblin.

references are quite often visualized by showing a radio or gramophone and thereby marked as a historical tune of the 1930s. In addition, Fassbinder uses a lot contemporary music (for example Janis Joplin's *Me and Bobby McGee* or the song *Radioactivity* by the German electro-pop band Kraftwerk) to dramatize or emotionalize specific situations. Music – both as instrument for the creation of a reality effect that suggests historical authenticity and as alienation effect in the Brechtian sense – is probably the strongest stylistic device used in the melodramatic staging of the story. According to the Swiss writer and literary critic Hermann Burger, the »requiem-like seesaw between polyphony and cacophony« that characterizes the soundscape of Fassbinder's film can be interpreted as the most successful translation of Döblin's technique of »filmic writing« into *audio*visual filmic storytelling.[109] As such, the film is an acoustic realization of the »style of simultaneity« for which Döblin's novel had been praised. For Burger, the film is less a visual than a sonic didactic play demonstrating the felicitous appropriation of Döblin's style of montage by Rainer Werner Fassbinder.[110]

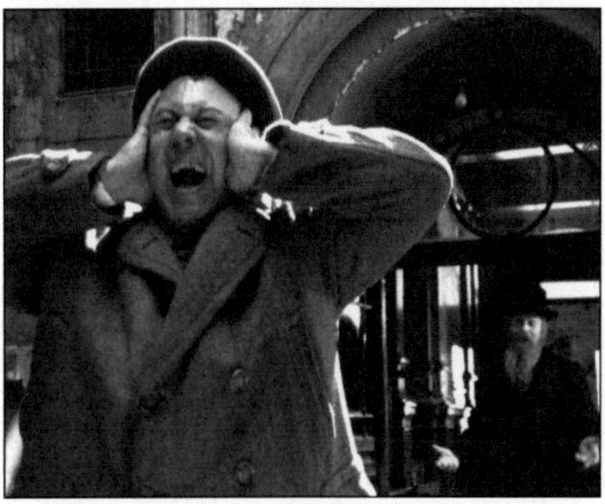

Figure 6: Still of the television series *Berlin Alexanderplatz* (1980) by Rainer Fassbinder: Biberkopf after his release from prison and holding his ears, trying to protect his head from the brutal intrusion of city noises.

The opening scene of the film (episode 1 entitled »Die Strafe beginnt« (The punishment commences – a literal quotation of the sentence following the opening paragraph of the novel) provides us with a telling example of how Fassbinder stages the city as a cacophony of noises, symbolizing both hectic big city life (as contrasted with the security and silence of Biberkopf's life in prison) and the inner tension and despair of the protagonist. While Biberkopf shows a glimpse of happiness when looking into the open sky of the prison yard that he had to cross in order to get to the main gate, he immediately

109 Burger, »Kein Platz für Alexanderplatz«, quoted in Sander 1998, 250.
110 Ibid.

shrinks back from the gate when confronted with the heavy traffic sounds coming from the street in front of the prison entrance. He even walks back into the yard, holding his ears and grimacing with pain. When he finally leaves the prison after some friendly suasion by the guardsman, Biberkopf runs blindly onto the street and is nearly run over by a car – despite repeated honking by the driver. In contrast to Jutzi's film version, where the confrontation between Biberkopf and the city is staged in the tram ride (using a rhythmic montage of shots and sounds to allegorize the rising inner tension and panic of the protagonist), Fassbinder deviates from the novel and does not refer to the tram ride at all. Instead of using the tram as symbol of the electrified mobility of big city life, Fassbinder stages the conflict between men and the city in a rather static form by using sound as key element in the dramatization of the narrative.

This rather static staging of *Berlin Alexanderplatz* by Fassbinder with its high theatrical production style (most scenes play in closed, slightly darkened rooms) and focus on the visualization of the inner monologues of Franz Biberkopf (with many close ups in order to catch the vivid and expressionist facial expressions of Günter Lamprecht), is without doubt a key feature of the film.[111] In disassociating the story from the concrete topography of the city of Berlin, Fassbinder condensates the existential orientation of a big city inhabitant to the mental state of an individual living in a cage, thereby underlining the desperate feeling of being locked-in.[112] Captured in his (mutilated) body, trapped by his social condition, and caught in his fantasies and desires – it is through this intimate portrait of Biberkopf's existential condition that Fassbinder visualizes, dramatizes and composes the conflict between individual and society in a modern urban environment. In this sense, Fassbinder's dramatization of the Biberkopf story comes very close to the montage-like narrative construction of Berlin as place of psychological estrangement and physical transit by Döblin. Both Döblin and Fassbinder construct their representations of Berlin and of the Alexanderplatz in particular as »psycho-topographies«, where

> city scenes are always conditioned as other scenes in the Freudian sense, distorted and displaced by desires and anxieties at both individual and collective level.[113]

111 The catalogue of the 2007 exhibition *Rainer Werner Fassbinder: Berlin Alexanderplatz*, curated by Klaus Biesenbach at the KW Institute for Contemporary Art in Berlin (March – Mai 2007), shows 560 (!) colour stills of the film offering an intriguing panorama of Fassbinder's visual imagination. See Biesenbach 2007.

112 Biesenbach 2007, 40-41.

113 Webber 2009, 261.

7. Conclusions

The challenge of the different media versions of *Berlin Alexanderplatz* analyzed in this chapter was simultaneously to represent and to evoke the feeling of the urban condition of men – and of one man in specific: Franz Biberkopf. Mobility, both in its cognitive and physical meaning, has probably served as the most crucial factor in visualizing and sonically evoking metropolitan life. In that respect, the idea of mobility inspired both the narrative construction of the novel and its filmic and radiophonic adaptations and served as a leitmotif for the characterization of Biberkopf's inner life. As Patrick Fortmann has argued, this double sense of mobility as both bodily and mental movement is a central characteristic of Döblin's novel and permeates the other adaptations as well:

> By contrasting the deficient narrative at the onset of the essay with new forms of representation in its body and by juxtaposing the sampling tourist with the true visitor seeking the whole of Berlin, Döblin not only underlines his call for a new aesthetics, one that is mindful of urban modernity, but also exposes different approaches to the metropolis and to the experiences it enables. At the center of these approaches is the transport of a subject through space, be it by tram, by bus or even on foot. Focusing on the interplay between movement and perception and oscillating between fragmented impressions and coherent gestalt, dismembered parts and organic whole, these approaches model at once the self in motion and the space traversed.[114]

The tension between »fragmented impressions and coherent gestalt« and »dismembered parts and organic whole« not only characterizes the mobile condition of men in a metropolitan environment, but also applies to the narrative and stylistic principles of the novel and its various adaptations. Döblin's *Kinostil* and *Tatsachenphantasie*, his montage-like combination of textual fragments originating from different media (newspaper, advertisements, radio announcements, musical lyrics, etc.) into impressionistic yet dense descriptions of Berlin as seething metropolis stages the city as a highly mediatized construct. But the many intermedial references not only serve as indicators of urban life, but exemplify a stylistic principle aiming at challenging the linear narrative of classic epic storytelling. The »deconstructive montage«-character of the novel aimed at reproducing the heterogeneity of the perception of reality at the level of aesthetic

114 Fortmann 2009, 126.

production.[115] In that respect, Döblin's *Berlin Alexanderplatz* put into praxis what the author theoretically had postulated in his »reflections on the novel« (»Bemerkungen zum Roman«, 1917) some twelve years earlier: »If a novel cannot be sliced into 10 pieces like an earthworm, and each part is able to move along, it is no good«.[116]

From a narrative point of view, neither the radio play nor the filmic adaptations by Jutzi and Fassbinder can claim such a radically new approach in storytelling. Although Döblin saw the innovative potential of the radio play as a new art form, technical limitations and aesthetic doubts frustrated his intentions to produce and transmit a radiophonic adaptation of his novel in 1930. But the 2007 version of the play, a contemporary production based on an accurate performance of the original manuscript written by Döblin, eventually shows the creative ideas and the eagerness to experiment with the new format – most convincingly demonstrated in the »humanization« of the *Fiat* and *Opel* cars. In contrast to the radio play, the filmic adaptations had the possibility of both visualizing and rendering audible the many intermedial references of the novel and to stage the city of Berlin as capital of visual and sonic spectacle. Yet they did so in quite different ways. While Jutzi's adaptation struggles with the customization of the novel into an emerging format of cinematic storytelling (sound film), Fassbinder's serial approach challenged classical forms of movie adaptations of literary works by a dramatic prolongation of the filmic narrative into a television series.

We do not aim at making any qualitative statement about the degree of »authenticity« of each of the adaptations in comparison to the original novel – a question that has largely dominated the intense public debate among critics and that continues to trigger the intellectual and scholarly discussion in literature and film studies. Instead, our comparative and intermedial approach aimed at analyzing and describing the complex relationship between the emergence of specific media products and their specific historical environment, and – more importantly – to look at the different ways of sonic representation and acoustic staging of Berlin in the novel, radio, film and television series. When revisiting the three forms of intermediality as outlined by Rajewski and introduced in the first paragraph of this chapter (intermediality as medial transposition, media combination or intermedial reference), two of them apply to the study presented here: first, the radio play, the film and the television series are of course adaptations of

115 Paech 1988, 130.

116 »Wenn ein Roman nicht wie ein Regenwurm in 10 Stücke geschnitten werden kann, und jeder Teil bewegt sich selbst, dann taugt er nicht«. See Döblin 1917, »Bemerkungen zum Roman«. According to Michael Opitz, Döblin's provocative thoughts on the novel have inspired Brecht's work on the »Verfremdungseffekt«. See Opitz 2011.

the novel and thereby a form of medial transposition; secondly, both the three adaptations and the novel provide ample evidence of intermedial references. Examples are the references to newspaper articles, posters, song lyrics or film titles in the novel, the musical references to popular Schlager songs and the playing of gramophone tunes in the radio play, the noisification of music or the acclamation of newspaper news in Jutzi's film, the period photographs and the frequent use of radio background sounds in Fassbinder's television series – to mention just a few of the many intermedial references.

As demonstrated, the transposition of the textual format of the novel into audio and audiovisual formats of storytelling in the radio play, film and television series was characterized by format-specific constraints: the complex story of a 400-page novel hardly fits into a 60-minute radio play or a 2-hour feature film. These format-specific constraints of the different media do not say anything about the aesthetic quality or narrative creativity of the adaptations as such, of course. The 2007 version of the radio play, Jutzi's early sound film adaptation and Fassbinder's monumental television series each demonstrate creative appropriation of the original text through exploration of the dramatic potential of the story in distinct ways. Yet all three adaptations share one important characteristic: while Berlin was the subject in Döblin's novel and Biberkopf its main object, the three adaptations turned Biberkopf into the main subject of storytelling through which Berlin was to be experienced (or not). This shift from Berlin as subject of storytelling in the novel to object of implicit or explicit urban reference in the adaptations has clearly affected the repertoire for the sonic representation of the city as well.

While the audiovisual medium of film and the acoustic medium of radio in theory seem to offer preferred means of staging the sonic experience of urban life, the choice of narrative conventions (mainly the focus on a single protagonist) and format-specific conventions (such as the length of a movie; the scheduling of the broadcasting of the series) has caused the city to fade from view. All three adaptations in fact »downgraded« Berlin to the status of supporting actor, even though the original novel aimed to be the city's megaphone. The most surprising and perhaps paradoxical finding of this comparative intermedial approach, then, is that the novel version emerges as the narrative format that is most explicit and eloquent in referring to the soundscape of Berlin by means of onomatopoetic language, the *O-Ton*-like literary reproduction of local dialect, and its many descriptions of noises, sounds or music. In addition to these explicit intermedial references, the novel's text implicitly evokes an acoustic environment of the city by playing with the sonic assumptions in the reader's mind. This rather unforeseen silencing of the city in the audio-visual adaptations of *Berlin Alexanderplatz* is perhaps bound to produce some noise in a future debate on intermedial historiography.

8. Sources

Aalbers, Jasper (2013 forthcoming): Echoes of the City. Staging Urban Soundscapes in Fiction Film, 1930-2000, Maastricht.

Arnheim, Rudolf (1936): Radio. The Art of Sound, London.

Barta, Peter (1996): Bely, Joyce, and Döblin. Peripatetics in the City Novel, Gainesville.

Barthes, Roland (1987): Introduction to the Structural Analysis of Narrative, in: Heath, Stephen (ed.), Image Music Text, London.

Bayerdörfer, Hans-Peter (1985), »Ghettokunst«. Meinetwegen, aber hundertprozentig echt. Alfred Döblins Begenungen mit dem Ostjudentum, in: Grimm, Gunter / Bayerdörfer, Hans-Peter (eds.): Im Zeihen Hiobs. Jüdische Schriftsteller und deutsche Literatur im 20. Jahrhundert, Königstein, 161-77.

Benjamin, Walter (1930): Berlin Alexanderplatz, in: Die Gesellschaft (June 1930).

Berbig, Roland et al. (2011): Berlins 19. Jahrhundert. Ein Metropolen-Kompedium, Berlin.

Bernhart, Toni (2008): Stadt hören. Auditive Wahrnehmung in Berlin Alexanderplatz von Alfred Döblin, in: Zeitschrift für Literaturwissenschaft und Linguistik 149, 55-67.

Biesenbach, Klaus (ed.) (2007): Rainer Werner Fassbinder. Berlin Alexanderplatz, Catalogue of the exhibition Fassbinder: Berlin Alexanderplatz – eine Ausstellung, KW Institute for Contemporary Art / Berlin (18 March – 13 May 2007), Mosel.

Biesenbach, Klaus (2007): Schwarze Bilder, in: Biesenbach, Klaus (ed.): Rainer Werner Fassbinder. Berlin Alexanderplatz, Mosel, 31-45.

Bird, Stephanie (2009): Nachum der Weise. On Storytelling, Eyes and Misunderstanding in Berlin Alexanderplatz, in: Davies, Steffan / Schonfield, Ernest (eds.): Alfred Döblin. Paradigms of Modernity, Berlin / New York, 245-260.

Bolter, Jay David / Grusin, Richard (1999): Remediation. Understanding New Media, Cambridge Mass.

Braunger, Amely (2006): Intermedialität und Urbanität im Medienwandel. Eine Untersuchung zur Darstellung der urbanen Lebenswelt in Hörspiel, Roman und Film am Beispiel von Berlin Alexanderplatz, Noderstedt.

Bräutigam, Thomas (2005): Hörspiel-Lexikon, Konstanz.

Brecht, Bertolt (1932/1967): Der Rundfunk als Kommunikationsapparat, in: Bertolt Brecht: Gesammelte Werke 18, Frankfurt am Main, 127-34.

Burger, Hermann (1987): Kein Platz für den Alexanderplatz. Zu Döblins Roman Berlin Alexanderplatz und zu Fassbinders Verfilmung, in: Als Autor auf der Stör, Frankfurt am Main, 186-89.

Davies, Steffan / Schonfield, Ernest (2009) (eds.): Alfred Döblin. Paradigms of Modernity, Berlin / New York.

Döblin, Alfred (1917): Bemerkungen zum Roman, in: Alfred Döblin. Schriften zur Ästhetik, Poetik und Literatur, herausgegeben von Erich Kleinschmidt (1989), Freiburg, 123-127.

Döblin, Alfred (1961): Berlin Alexanderplatz. Die Geschichte vom Franz Biberkopf, Freiburg im Breisgau.

Döblin, Alfred (1961): Mein Buch »Berlin Alexanderplatz«, in: A. Döblin, Berlin Alexanderplatz. Die Geschichte vom Franz Biberkopf, Freiburg im Breisgau, 505-07.

Döblin, Alfred (1997): Berlin Alexanderplatz. The Story of Franz Biberkopf, New York.

Elsaesser, Thomas (2010): The Fassbinder Foundation, http://www.fassbinderfoundation.de/node.php/en/leben.

Fassbinder, Rainer Werner (2006): Die Städte des Menschen und seine Seele. Einige ungeordnete Gedanken zu Alfred Döblins Roman Berlin Alexanderplatz, originally published in Die Zeit (14 March 1980), reprinted in: Fassbinder. Berlin Alexanderplatz remastered, booklet of the Arthaus DVD edition (2006) of the film, 5-15.

Fickers, Andreas (2012): The Emergence of Television as a Conservative Media Revolution. Historicising a Process of Remediation in the Post-war Western European Mass Media Ensemble', in: Journal of Modern European History 10.1, 49-75.

Flesch, Hans (1928): Hörspiel – Film – Schallplatte. Referat auf der ersten Tagung des Programmrats in Wiesbaden am 5.-6. Juni 1928, in: Rundfunk-Jahrbuch (1931), 31-63.

Fortmann, Patrick (2009): Topographies of Transport. The Openings of Alfred Döblin's *Berlin Alexanderplatz*, in: Reinhard Zachau (ed.): Topography and Literature. Berlin and Modernism, Göttingen, 125-39.

Fritzsche, Peter (1996): Reading Berlin. Cambridge Mass.

Gilfillan, Daniel (2009): Pieces of Sound. German Experimental Radio, Minneapolis.

Goebel, Rolf (2009): Media Competition. Ruttmann's Berlin: die Symphonie der Großstadt and Hessel's Ein Flaneur in Berlin, in: Zachau, Reinhard (ed.): Topography and Literature. Berlin and Modernism, Göttingen, 111-24.

Gomery, J. Douglas (1976): The coming of talkies. Invention, innovation, and diffusion, Madison, Wis.

Göttlich, Fritz (2007): Die Finsternis eines Meisterwerks, in: Fassbinder. Berlin Alexanderplatz remastered, booklet of the Arthaus DVD edition (2006) of the film, 30-32.

Graham, Allen (2011): Intertextuality, London.

Grob, Norbert (2003): Phantasma Stadt / Phantasma Strasse. Zur großstädtischen Aura in Filmen der zwanziger und frühen dreißiger Jahre, in: Koebner, Thomas (ed.): Diesseits der »Dämonischen Leinwand«. Neue Perspektiven auf das späte Weimarer Kino, Munich.

Haag, Achim (1991): Fassbinder verfilmt *Berlin Alexanderplatz*. Bilder und Töne jenseits ihrer Vorlage: Wider eine Dogmatik der Literaturverfilmung, in: Stauffacher, Werner (ed.): Internationale Alfred-Döblin-Kolloquien Münster *1989*, Bern, 302-10.

Hake, Sabine (1994): Urban Paranoia in Alfred Döblin's *Berlin Alexanderplatz*, in: The German Quarterly 67.3, 347-68.

Hake, Sabine (2008): German National Cinema, 2nd Edition, London / New York.

imDb: Gier, Trivia, http://www.imdb.com/title/tt0015881/trivia.

Jelavich, Peter (2006): Berlin Alexanderplatz. Radio, Film, and the Death of Weimar Culture, Berkeley.

Kaes, Anton (1978): Kino-Debatte. Literatur und Film 1909-1929, Tübingen / Munich.

Klotz, Volker (1987): Die erzählte Stadt. Ein Sujet als Herausforderung des Romans von Lesage bis Döblin, Munich / Wien.

Lorenz, Juliane (2006): Retour zum »Alexanderplatz«, in: Fassbinder. Berlin Alexanderplatz remastered, booklet of the Arthaus DVD edition (2006) of the film, 16-27.

Melcher, Andrea (1996): Vom Schriftsteller zum Sprachsteller? Alfred Döblins Auseinandersetzungen mit Film und Rundfunk (1909-1932), Frankfurt am Main.

Mühsam, Erich (1984): Franz Biberkopf und der Rundfunk, in: Schneider, I. (ed.): Radio-Kultur in der Weimarer Republik. Eine Dokumentation. Tübingen, 131-33.

Müller, Eggo (1988): Das Verhältnis von Literatur und Film. Eine Untersuchung von Alfred Döblins Schriften zu Roman und Film und eine Interpretation von »Wadzeks Kampf mit der Dampfmaschine« und »Berlin Alexanderplatz«, unpublished MA thesis, Free University Berlin.

Müller, Eggo (1992): Adaption als Medienreflexion. Das Drehbuch zu Phil Jutzis Berlin Alexanderplatz von Alfred Döblin und Hans Wilhelm, in: Schwartz, Alexander (ed.): Das Drehbuch. Geschichte, Theorie, Praxis, München (Münchner Beiträge zur Filmphilologie 5).

Müller, Jürgen (2008): Intermedialität und Medienhistoriographie, in: Paech, Joachim / Schröter, Jens (eds.), Intermedialität Analog / Digital. Theorien – Methoden – Analysen, Munich, 31-46.

Muschg, Walter (1961): Nachwort des Herausgebers, in: Döblin, A., Berlin Alexanderplatz. Die Geschichte vom Franz Biberkopf, Freiburg im Breisgau, 509-28.

Opitz, Michael (2011/10/04): book review of the Döblin biography by Wilfried F. Schoeller: Alfred Döblin – Eine Biografie, Deutschlandradio Kultur, http://www.dradio.de/dkultur/sendungen/kritik/1569502/.

Paech, Joachim (1988), Literatur und Film, Stuttgart.

Prangel, Matthias (ed.) (1975): Materialien zu Alfred Döblins »Berlin Alexanderplatz«, Frankfurt am Main.

Rajewsky, Irina (2002): Intermedialität, Tübingen.

Rajewsky, Irina (2005): Intermediality, Intertextuality, and Remediation. A Literary Perspective on Intermediality, in: Intermédialités 6, 43-64.

Rehhahn, Yvonne (1996): Nur der Veränderte Autor Kann den Film Verändern, in: Berlin-Alexanderplatz. Drehbuch von Alfred Döblin und Hans Wilhelm zu Phil Jutzi's Film von 1931. Munich.

Sander, Gabriele (1998): Erläuterungen und Dokumente. Alfred Döblin: Berlin Alexanderplatz, Stuttgart.

Schoeller, Wilfried F. (2011): Alfred Döblin – Eine Biografie, Munich.

Schwitzke, Heinz (1976): Die Geschichte vom Franz Biberkopf. Hörspiel nach dem Roman Berlin Alexanderplatz, Stuttgart.

Shail, Andrew (2010): Intermediality. Disciplinary Flux or Formalist Retrenchment?, in: *Early Popular Visual Culture* 8.1, 3-15.

Sontag, Susan (1983/2007): Ein Roman wird Film. Fassbinders Berlin Alexanderplatz, in: Biesenbach, Klaus (ed.) (2007): Rainer Werner Fassbinder. Berlin Alexanderplatz, Catalogue of the Exhibition: Fassbinder. Berlin Alexanderplatz – eine Ausstellung, KW Institute for Contemporary Art / Berlin (18 March-13 May 2007), Mosel, 48-53.

Stoffels, Ludwig (1997): Kunst und Technik, in: Leonhard, Joachim-Felix (ed.): Programmgeschichte des Hörfunks in der Weimarer Republik 2, Munich, 682-724.

Uricchio, William (1982): Ruttman's »Berlin« and the City Film to 1930. PhD Thesis, New York University.

Vogt, Guntram (2001): Die Stadt im Kino. Deutsche Spielfilme 1900-2000. Marburg.

Ward, Janet (2001): Weimar Surfaces. Urban Visual Culture in 1920s Germany, Berkeley / Los Angeles.

Webber, Andrew J. (2009), Other Places. Döblin's Berlin Topographies, in: Davis, S. / Schonfield, E. (eds.): Alfred Döblin. Paradigms of Modernity, Berlin / New York, 261-75.

Wikipedia: http://de.wikipedia.org/wiki/Einwohnerentwicklung_von_Berlin.

Williams, Raymond (1987): Country and City in the Modern Novel. W.D. Thomas Memorial Lecture, Swansea.

Wurlitzer, Steve (2007): Electric Sounds. Technological Change and the Rise of Corporate Mass Media, New York.

The Chirping of a Little Bird

Some (Film) Theoretical Reflections

Patricia Pisters

1. Introduction

Franz Biberkopf holding his hands over his ears to protect him from the city noise when he leaves the quietness of the prison walls has become the iconic sound-image of Rainer Werner Fassbinder's version of *Berlin Alexanderplatz*. As the authors of »Sounds Familiar« (this volume) have shown, Fassbinder's epic translation of Döblin's book does not make extended use of the topology of the city, but the overwhelming traffic noise has indeed become a familiar sound trope of Berlin as exemplary modern city of the early twentieth century. Focusing on the mental state of Biberkopf, other sounds of the city are not immediately obvious in Fassbinder's *Alexanderplatz*. Therefore it is interesting to return to the city of Berlin and make a few (film) theoretical sound excursions to suggest some further reflections on Fassbinder's mental soundtrack and the ways in which it can be related to the soundscape of the city. In discussing the difference between expressionistic and realistic sound effects, paying attention to the ways in which the sound track is related to gender differences and by relating sound to its powers to create and invade spaces, I will propose some additional intermedial observations to the ones already put forward in »Sounds Familiar«. Moreover, I suggest that methodologically we remember sounds of the urban past *trans*medially in the sense that with every new technologically mediated soundscape, we add a layer to our aural memory that is transported through the different media. Such as the simple sound of a chirping little bird.

2. Returning to Alexanderplatz

For the television programme *In Europa* (VPRO 2007) Günther Lamprecht, who legendarily played Biberkopf in Fassbinder's epic television series, returns to Alexanderplatz of the year 2007. Standing in the middle of the square, closing his eyes he goes seventy years back in time and imagines the sounds of this place when he grew up: horses and carriages, breweries that carried their barrels; underneath the station the voices of the vendors of the big food market. These were the sounds of Berlin of his youth, Alexanderplatz of the 1930s. Opening his eyes, Lamprecht continues, »Now, I hear mass tourism«. In this particular episode of *In Europa* about Berlin in the 1930s, historian Geert Mak recalls in voice-over Fassbinder's Biberkopf as the symbol of the moral downfall of mankind, caught between the anarchistic decadence of the Weimar republic and the emergence of Nazi ideology. Compared to the rare city sounds in Fassbinder's film, Lamprecht's own memories of the city itself are rich in sound qualities. Still on Alexanderplatz in 2007, Lamprecht recalls how in 1945 he witnessed the end of the war at the exact same spot when the whole square was just rubble and steel. When fourteen-year-old Lamprecht came out a bunker, hands in the air, a drunken Russian soldier told him: »Wojna kaput, wojna kaput«, the war is over. »After two weeks of constant noise of drumfire and bombings, it was suddenly silent«, Lamprecht remembers.

In contrast to the observations in *Shifting Sounds* earlier in this volume, where it was described how silence in Amsterdam gradually became a sign of war in the early 1940s, Lamprecht remembers silence as the sound of peace. But there was one other little sound that he vividly remembers in connection to this silence: from underneath the debris he suddenly heard a very soft squeeking sound. And then he saw a little bird, a sparrow. As if these little chirping sounds had been hiding under the ruins and the blanket of violent war noise and only now dared to appear again. While Lamprecht sighs that with this sound the war was definitively over, he directs our attention to a sparrow's twittering on present-day Alexanderplatz which at the very moment he talks about his sonic memory enters the frame and the sound space of the present-day camera. It is a magical moment, where past and present collide through an aural sign. And where a little bird's chirp becomes part of a city soundscape with huge historical significance. I will return to the squeeking sound of a little bird later on, when paying attention again to Fassbinder's *Alexanderplatz*. In most of this expansion, however, I turn to some sonic concepts in film theory and philosophy that might be brought in dialogue with the findings in the previous chapter, especially in comparing the different audio-visual adaptations of Döblin's novel.

3. Realistic versus Expressionistic Uses of Sound

As mentioned the first thing that is remarkable in Fassbinder's *Berlin Alexanderplatz* is that frequently the soundtrack of the city gives way to the internal world of Biberkopf. A theoretical distinction that can be made in this respect is the one between expressionist and realistic uses of sound. It is important to note that, as emphasized by Michel Chion in his *Audio-Vision*, there is no »natural and pre-existing harmony between image and sound«.[1] We relate images and sounds with our brain, and filmmakers can experiment with these relations, but there is no law that makes images and sounds go naturally together. Nevertheless, since the official introduction of sound film in 1927, the soundtrack has been constructed largely in function of realistic representation and the predominant mirroring of sound to vision: what we see is what we hear. In Phil Jutzi's film version of *Berlin Alexanderplatz* (1931) the soundtrack is realistic. For instance, when Biberkopf leaves Tegel prison at the beginning of the film, both the music and the sound effects of tram bells and car horns get louder and more agitated when Biberkopf enters more deeply into the crowdedness of the city centre. The sound thus follows realistic conventions: we travel with Biberkopf to the city and become gradually overwhelmed by the images and sounds: what he perceives is what we perceive.

In contrast to this realistic use of sound that is always externally (objectively) motivated, filmmakers have also experimented with expressionist soundtracks, which are more internally motivated and could be described as »what the character feels is what we hear«. Since his first sound film *Blackmail* (1929), famous for the sound close-up of the word »knife« that translated the obsessive thoughts of the heroine of the film, Alfred Hitchcock has experimented with different forms of internal sounds that express a character's thoughts, fantasies or emotions. Two more examples of early sound experiments in Hitchcock's films are the internal monologue in *Murder!* (1930) and another obsessive state of mind of a character which transforms music into noise in *Secret Agent* (1936).[2] While the interplay between image and sound indicates the expressionistic or realistic status of the sound (such as determining the source of the sound in the image), the quality of the sound itself (pitch, volume, direction) may indicate its status as expressionistic or realistic sound. Clearly the volume of the city cacophony hurting Biberkopf's ears in the first scene of Fassbinder's *Alexanderplatz* translates the sounds of the city in an expressionistic way: Biberkopf is terribly afraid to enter the chaos of

1 Chion 1994, xvii.

2 Pisters 2007, 169-73.

urban life. So even while just outside the prison the city is not very audible yet, we do hear its overwhelming sounds as the expression of an internal experience.
Could these sounds be representative of the actual city, or are there ways to translate the expressive inner experiences of Biberkopf to what is going on in the big city? On a comparative level, a more extended argument perhaps could be made to see how the expressionistic use of sound in Fassbinder's version relates to a more realistic approach of the sounds of the city in the radio play and the first filmic interpretation of Döblin's novel. Do we hear different cities in these various versions?

4. Acousmatic Sounds and Embodied Voices

Another theoretical observation made by Chion concerns the particular use of sound that does not have a visible source in the image track that are not necessarily expressionistic or psychological sounds. This type of sound, such as a non-diegetic voice-over, is called »acousmatic« sound.[3] »Acousmatic« is an old Greek term used by Pythagoras to indicate that masters were hidden behind a curtain when teaching their pupils. They were not visible; only their voice was audible, which increased their power, as if they were speaking with the voice of God. Chion has a special term for filmic characters that are powerful because of their acousmatic presentation: the »acousmêtre« (a contraction of acousmatic and *être en maître*). The most famous examples are the wizard in *The Wizard of Oz* (Victor Fleming, 1939) and Dr. Baum/ Dr. Mabuse in *The Testament of Dr. Mabuse* (Fritz Lang, 1933): as long as they are not visible they are omnipotent. As Kaja Silverman, referring to Chion, has argued, in classic cinema also the all-knowing voice-over is reserved mostly for male characters.[4] If there is a female voiceover, it is an embodied one, the voice of an actress whom we can see in the image while she is speaking in voice-over. And this embodiment gives her less authority over the images, over the narration.
Another difference in relation to the voice that Silverman observes is the opposition between the maternal voice and the paternal word, and it is mostly the word – the discursive and signifying power – that wins this battle, at least in classic cinema. In this respect Silverman also discusses Chion's argument that cinema is »a machine made in order to deliver a cry from the female voice«.[5] Obviously this does not mean that

3 Chion 1994, 129.

4 Silverman 1988.

5 Chion in Silverman 1988, 77.

women do not speak in classical cinema. In »Shifting Sounds« the example of *Out of the Clouds* (1955) is given. Here the female owner of a pub explains the sounds of London. But these are soothing words to familiarize and reassure in a motherly way strangers with a new environment, making the city a *home*. What is demanded from women on a more profound level in cinema, Silverman argues, is an involuntary sound: the scream, the cry, or possibly a (non-discursive) melody. Ranging from Fay Wray's scream in *King Kong* (Merian Cooper, 1933) to the »perfect scream« in *Blow Out* (Brian de Palma, 1981), cinema has indeed produced a whole range of female screams. Male screams are hard to find in classic Hollywood cinema. The female voice is embodied within the diegetic world and has often much less discursive power. Even vocal resistance takes place in this non-discursive level. In this respect the counter-sound of the female cry is laughter. In her book *The Unruly Woman* Kathleen Rowe has demonstrated how laughing women are considered as improper and transgressing many conventional gender borders of decent behaviour.[6] And in this respect it is interesting to see that in both Jutzi's film and in Fassbinder's television play there are some remarkable, almost uncanny scenes of laughing women.

So the quality of the voices, their acousmatic status and their (gendered) relation to the utterance of words or just sounds, screams or laughter would therefore offer another possible perspective on the different soundscapes of *Berlin Alexanderplatz*. The radio play, for instance, although an acousmatic medium in itself, allows interesting reflections on questions of embodiment (»a voice comes from a throat«), power and perhaps also gender relations. The female voices in the 2007 version of the play are incredibly soft and childlike. And in Fassbinder's television epic, it is interesting to note that Biberkopf's traumatic memory of him murdering his wife Ida (for which he served a prison sentence) is repeated in six flashbacks. The images of the flashback return in exactly the same way, but each version is overlaid with an acousmatic voice, which relates Biberkopf's affective and personal memory to a different set of objective facts: ranging from a scientific observation of the physical laws that explain what happened to Ida's rib case, to scenes from the city (a horse falling into a pit, the collapse of a huge building), political and historical events and news of the day.[7] The voice is Fassbinder's own male voice, which claims some authority over the objective facts laid over the subjective memory. This is another way in which the internal and external meet in an audio-visual encounter. Or as Biberkopf states at the beginning of the Fassbinder's *Alexanderplatz*, of the encounter of »the city, the world and me«.

6 Rowe 1995.

7 Del Rio 2009, 73-96.

Also the voices of the women in Berlin Alexanderplatz might be analyzed further and in a comparative way. Do they have a different status in the radio play than in the film or the television play? With what authority do they speak, when and how do they sing, laugh, are silent or scream? One of the most interesting scenes in this respect is Mieze's scream in Episode 11 of Fassbinder's *Alexanderplatz*. In this scene Biberkopf has Reinhold hide in bed in their apartment to show him braggingly Mieze's devotion. But when Mieze arrives she tells Franz she is in love with another man [...] In response to Franz's rage she »delivers a cry« to the point that Franz throws himself on her and starts to scream with her before assaulting her. There is a brief but very powerfully sonic moment when in their screams, their despairs meet, before the violence takes over again. I will return to this scene at the end of these brief reflections. Here I have pointed to the possible gender relations that we can hear on the different soundtracks of *Berlin Alexanderplatz*.

5. Terriorializing and Deterritorializing Effects of Sound

A final sonic conceptual thought that I would like to bring into the debate concerns the idea of territorializing and deterritorializing effects of the soundscape. Here I will refer to *A Thousand Plateaus*, where Gilles Deleuze and Félix Guattari write extensively about sounds, music and the voice.[8] Before going back to soundtracks as such, I raise some points that Deleuze and Guattari introduce in relation to music: the territtorializing refrain and the deterritorializing »becoming-music«. An important aspect of music is its power to create territories. According to Deleuze and Guattari, the refrain is territorial and has the function of creating »safe havens« in the chaos of the world: a child comforts itself in the dark by singing softly a nursery rhyme; a bird launches into its refrain marking its territory.[9] Every household, too, is an aurally marked territory. Homes are created by sound walls: a radio that is playing, singing and speaking voices, the sound of the washing machine – they are all the aural markers of a familiar territory. Countries are created by national anthems. Sound has even much stronger capacities to territorialize than sight, Deleuze and Guattari argue. As they concisely put it, »Flags can do nothing without trumpets«.[10] These are the territorializing powers of sound and music, of the refrain.

8 Deleuze / Guattari 1988, 299-350. See also Pisters 2003, 175-215.

9 Deleuze / Guattari 1988, 299-300.

10 Ibid., 348.

But the territory, the home or country can also be invaded, opened up by sound:

> One opens the circle a crack, opens it all the way, lets someone in, calls someone, or else goes out oneself, launches forth. [...] to join with the World, or melt with it.[11]

Sound tends to dissolve and connects with other elements easily. According to Deleuze and Guattari, sound is considered in its potential force to enable all kinds of becomings: sounds, music, noises, singing open up to other spaces: a singing voice enters into the territory of a bird, it enters into a »becoming-bird«. One can think of the singing opera voices in Frank Darabont's film *The Shawshank Redemption* (1994) where the most transgressive act of one of the prisoners in the Shawshank prison is the playing of an opera record. By hearing the voices of the opera singers, the prisoners were freer than if they had left the prison walls behind them. These are the deterritorializing forces of sound, of »becoming-music«.

So according to Deleuze and Guattari, territorializing, deterritorializing and reterritorializing forces are to be taken into consideration when we talk about sound.[12] Walter Murch, in his foreword to Chion's Audio-Vision also points to the territorial forces of sound:

> There were of course many more significant reasons for the rise of the Great Dictators in the twenties and thirties, and it is true that the silent film had sometimes been used to rally people around the flag, but it is nonetheless chilling to recall that Hitler's ascension to power marched in lockstep with the successful development of the talking film.[13]

The power of sound is, of course, even more obvious when we think of the power of Hitler's voice through the radio. Chion's concept of the »acousmêtre« now can be explained in terms of territorializing forces of sound.

Listening to *Berlin Alexanderplatz* according to these (de)territorializing characteristics, it would be interesting to analyse how, in the different versions of *Berlin Alexanderplatz*, sound creates the sense of a home – related to interior spaces as well as the city as a whole – and at which moments the territorial forces are opened up with deter-

11 Ibid., 311.

12 Ibid., 348.

13 Murch in Chion 1994, xi.

ritorializing aspects. One clear example can be given from the epilogue of Fassbinder's *Alexanderplatz*. After the murder of Mieze, Biberkopf ends up in a mental hospital. Many scenes translate his state of mind in surreal images full of »becoming-animal«, often displayed through bodily gesture and posture. In one scene Franz and Mieze are slaughtered like cows in an abattoir. In another scene Franz drinks milk from a saucer on the floor like a cat. Here the music and voice of Janis Joplin's »Me and Bobby McGee« indicates the deterritorializing effects of becoming-animal emphasized by the lyrics (»Freedom is just another word for nothing to lose«). In the final scene of Fassbinder's version, Biberkopf is *domesticated*: as an assistant gatekeeper at a factory he is alert at his job site, but no longer transpierced by the sounds of the city. Instead, *Berlin Alexanderplatz* ends with a folk song / anthem while the country marches into the war.

6. The Scream, the City, the Bird

At the end of this short reflection on some theoretical aspects of sound in film I would like to return to the above-described scene of Mieze screaming. Her screaming is so penetrating and loud that it really has deterritorializing powers. As Biberkopf exclaims: »Mieze, you scream the house down!«. By screaming she not only »breaks down« the walls of the house that is no longer a safe place, but it is as if she lets in the chaotic and dangerous forces of the city on the verge of a historical transition. But even while she screams her lungs out, there is still one other little sound audible: a bird's thin squeaky sound, Mieze's canary, remains audible throughout the whole scene. A soft chirping sound under the rubbles of a sonically torn-down house. After having listened to Lamprecht's sound memories of Alexanderplatz at the end of the war, the chirping of the sparrow emerging from the debris, one cannot avoid connecting the two sounds. The sound of a little bird in *Berlin Alexanderplatz* is about to die, only to be revived many years later, after the war trumpets have silenced. And hearing a bird chirping in my city garden in Amsterdam does not just sound familiar – it opens layers of sonic memories of a mediated urban past.

7. Sources

Chion, Michel (1994): Audio-Vision. Sound on Screen. Trans. Claudia Gorbman, New York.

Deleuze, Gilles / Guattari, Félix (1988): A Thousand Plateaus. Capitalism and Schizophrenia. Trans. Brain Massumi, London.

Pisters, Patricia (2003): The Matrix of Visual Culture. Working with Deleuze in Film Theory, Stanford.

Pisters, Patricia (2007): Lessen van Hitchcock. Een Inleiding in Mediatheorie. Third revised edition, Amsterdam.

Rowe, Kathleen (1995): The Unruly Woman. Gender and the Genres of Laughter, Austin.

Silverman, Kaja (1988): The Acoustic Mirror. The Female Voice in Psychoanalysis and Cinema, Bloomington and Indianapolis.

Del Rio, Elena (2009): Violently Oscillating. Science, Repetition and Affective Transmutation in Fassbinder's Berlin Alexanderplatz, in: Deleuze Studies 3.1, 73-96.

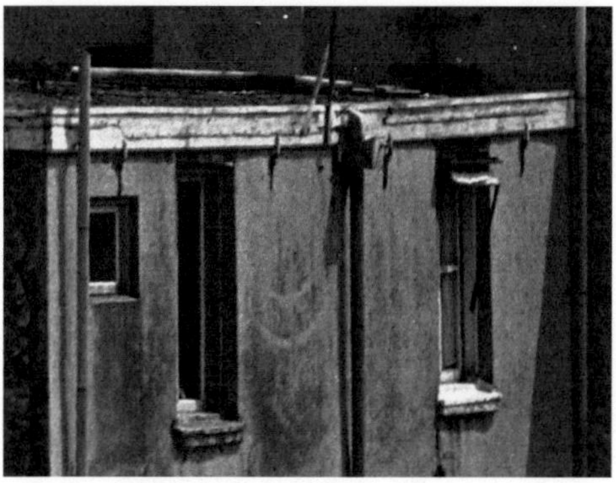

Sonic Artefacts

Reality Codes of Urbanity in Early German Radio Documentary

Carolyn Birdsall

Cities can be reflected in the camera; cities come into being on radio.[1]

The purely aural world [...] is relatively poor in documentary qualities.[2]

1. Introduction

Despite assertions that radio was a *blind* medium, early German radio practitioners sought to convey the sounds and sights of urban environments, whether from their own cities or faraway places.[3] City life was marked by the effects of industrialization and urbanization, and the intense concentration of activity in the city – with its pulsating crowds and traffic – was claimed to cause an overstimulation of the senses. In Germany, numerous commentators defined urban experience in terms of the intensification of *visual* stimuli, which was explicitly commented upon in photography, film, and modernist painting and literature.[4] At the same time, sounds were also asserted as definitive markers of life in the modern metropolis.[5] Nonetheless, there was not a fully-fledged practice for documenting the sonic dimensions of cityscapes. While modern music cited

1 »Städte spiegeln sich in der Kamera – Städte erstehen im Funk«. Marek 1930, 73.

2 Arnheim 1933/1972, 279.

3 Cf. ibid.

4 See, for instance, Simmel 1903/1997.

5 Cf. Lessing 1908, 1909. For a comparative account, see Bijsterveld 2001.

characteristic urban sounds, sometimes using horns and sirens in performance, technologies like the gramophone were primarily employed for recording speech and musical performance rather than urban soundscapes.[6]

In contrast to the previous chapters in this volume, which mainly focus on the representation of urban soundscapes in fictional genres, this chapter takes documentary as its starting point. Documentary portrayals of the city are usually considered in terms of visual or textual strategies; I will instead examine documentary impulses in early German radio broadcasting from 1923 onwards. The first decade of radio broadcasting coincided with the rise of the modern documentary film, which in numerous accounts is cited in reference to Robert Flaherty's films *Nanook of the North* (1922) and *Moana* (1926). The contribution of radio to the historical documentary project, however, has been significantly under-researched.[7] This may be in part due to the uneven development of *documentary* within early radio, which did not gain institutional and critical status in any way comparable to the documentary film movement

One of the challenges to examining documentary sound in radio portrayals of the urban is the limited availability of archival recordings. The *sonic artefacts* that will be discussed in this chapter are still accessible because these programmes were pre-recorded and recorded on shellac discs, unlike the majority of live programming in the 1920s. While these recordings may form part of an audiovisual heritage of urban soundscapes, this chapter does not partake in the discourse that mourns the lost or ephemeral nature of the sound archive. As radio scholar Josephine Dolan emphatically argues, written sources are also essential to the study of (British) broadcast history:

> The transmitted voices that are the focus of the sound archive can not be isolated from the voices of the written policy statements about audition, selection criteria, scripts and performance standards that are anterior to the moment of transmission. The moment of transmission [...] is a highly orchestrated production that is fully located within the complex relationships between the BBC, its personnel, its imagined audience and its empirical audience.[8]

6 On the technological restrictions posed on artistic projects to use phonographs for outdoor recordings, see Kahn 1994. When (outdoor) phonographic recordings were conducted, it was often for the ethnographic study of non-Western people. Cf. Stangl 2000, Ames 2007, Lange 2007.

7 For several notable exceptions, see Scannell 1986, Madsen 2009, Hendy 2004. For the development of documentary forms in the context of American radio, see, for instance, Loviglio 2002, as well as Bluem 1965, 60-72, Barnouw 1978, 28-37, McDonald 1979, 281-325, Lichty and Bohn 1974, 458-62.

8 See Dolan 2003, 69.

In other words, it is crucial that the investigation of documentary sound in German radio portrayals of the urban does not overprivilege archival recordings as preferable or superior to written and printed sources. With Dolan's comments in mind, the analysis here will also refer to programmes that were not recorded or preserved, and will engage printed archival documentation to situate the selected programmes in terms of the aesthetic choices, technological possibilities and institutional context of their production.

In what follows, I will begin with some conceptual groundwork concerning the notion of documentary. This requires attention to debates about indexicality in relation to photography and cinema, as well as discourses of documentary realism and the modernist preoccupation with urban subject matter. When adopting the documentary concept for radio history, it is important to acknowledge that this is a retrospective labelling, since »documentary« was not necessarily a unified field of production and reception at the time. Likewise, attempts to capture the *sounds of the real* predate radio broadcasting, and a prehistory could be traced in telegraphy, telephony, gramophone and early wireless practices. Having introduced key concepts for this investigation I will give further reasoning on my decision to focus on German radio broadcasting, which was launched in late 1923. I stress the particular background to state-governed radio broadcasting and the significant engagement of German intellectuals and artists with the medium in the decade following 1923. In the analysis, I will identify three main currents in documentary sound portrayals of the urban in Weimar-era radio. Firstly, and primarily in the early years of radio broadcasting, the project to document the urban took the form of studio-produced city portraits (*Hörbilder*), and, later, more ambitious outdoor broadcasts with the aid of sound recording and editing. This first analysis will acknowledge the relative fluidity between fiction and nonfiction categories in early radio explorations of the urban, which took place prior to the formal institutional separation between drama and news departments. In addition to modernist documentary concepts and sound experiments, I will, secondly, investigate the notion of actuality within early radio, and the development of conventions of news and live event broadcasting. In this case, I will pay particular attention to the rise of radio reportage with the aid of microphones, vocal presentation, outdoor location recording (*Originalton*) and gramophone prerecording, as important conditions in the historical invention of media *reality codes*.[9] In addition to case studies of *documentary* and *actuality*, I will investigate a third, related aspect of *authenticity*, by considering the recourse to particular uses of spoken voice (dialect), local music repertoire and urban narratives, in particular during the years of crisis be-

9 *Originalton* has a range of possible English translations including original sound, wild sound, actuality or ambiance. See Madsen 2009.

tween 1930 and 1933. In these three analytical sections I seek to draw out three main sonic strategies in portrayals of the urban: the *wandering* microphone (an imaginative, explorative mode), the *bird's eye* view (often providing a fixed, aerial perspective) and, finally, the transposition of urban documentary techniques to convey the nature of the region or the nation as a whole.

2. Documentary Realism, Urbanity and Radio Broadcasting

There are two common observations in scholarly criticism about the historical development of documentary. The first is that early cinema exhibition – such as by the Lumière Brothers and Skladanowsky brothers in 1895-96 – featured short recordings of city scenes, such as factory workers, trains in motion and inner city traffic. As Guntram Vogt notes, already by 1905, the fascination with the city had led to film sequences »moving over rooftops, along city walls, through courtyards, on the river and out of the city, through to open spaces«.[10] In addition to this preoccupation with urban scenes, a second observation concerns the shared origins of fiction and nonfiction film in early cinema, much of which »casually blended the staged and unstaged, actors and non-actors, fact and fiction«.[11] Indeed, prior to the establishment of narrative cinema as a dominant mode in the late 1910s, the actuality film was a prevalent feature of early cinema exhibition, with the popularity of topical themes, street scenes, exotic scenes or travelogues already suggesting the appeal of *real sights* for audiences. Both of these aspects – the city as subject matter and the *blurred boundary* of cinema history prior to a fiction-nonfiction divide – provide impetus for this section, which will focus on documentary film discourse, notions of documentary realism and the portrayal of urban environment. In terms of sound, the most immediate observation about 1920s documentary film is that it was *silent*, with its truth claims linked to the photographic.[12] The concepts of the document and documentary as associated with (written) evidence is consistent with both the historical rise of media of mechanical reproduction (photography, phonography, cinema) and the professional field of history in the nineteenth century.[13] This

10 »...über Dächer, entlang an Außenmauern, durch Hinterhöfe, hinaus der Stadt auf den Fluß, aufs offene Land«. Vogt 2001, 123.

11 Nichols 1995, n.p.

12 Cf. Altman 2004.

13 Cf. Rosen 1993.

simultaneous development gave rise to notions of knowledge acquisition by observing the traces of the past, as evidenced by authentic historical documentation, written or otherwise. A similar investment in documentary evidence was present in the emergent fields of ethnography and anthropology, with figures like Franz Boas promoting phonographic recordings for fieldwork, since it

> provided a form of documentation directly apprehensible to the senses, presumably uncontaminated by the observer's inevitable categorical and perceptual biases.[14]

In other words, indexical media technologies were increasingly affirmed as facilitating the collection of *pure data* for the task of capturing reality. This rhetoric of science, in turn, was also intrinsic to early discourses of documentary film, since photography's ability to reproduce the real was perceived as similar to that of scientific tools (such as the thermometer).[15]

The photographic dimension to cinema has traditionally offered a marker that distinguishes it from other documentary or realist strategies in literature or painting.[16] In the particular case of documentary film, realist theorizations have been countered with rereadings of documentary history that foreground the poetic and expressive. The silent *city films* of the 1920s are often cited as a case in point for both formal experimentation and social commentary. The most well-known European examples of this trans-national genre include Alberto Cavalcanti's *Rien que les heures* (1926) and Jean Vigo's *Á Propos de Nice* (1930), and feature-length films by Walther Ruttmann (*Berlin: Sinfonie der Großstadt*, 1927) and Dziga Vertov (*Man with a Movie Camera*, 1929). The avant-garde genre of the *city film* is usually identified in terms of the emergence of a new visual language that tried to capture the distinct modern experience of urban life, often with the aid of montage and the metaphor of the symphony.

For many, there was a sense in which synchronized sound was perceived as interfering with cinema's modernist project, and silent documentaries continued in different forms until at least the 1960s.[17] Some of the initial setbacks with sound film included limited movement of cameras, which were primarily confined to the studio, with post-production sound techniques only possible after 1932. Following the ascension of sound

14 Brady 1999, 69.

15 Winston 1993.

16 For a good overview of debates on realism in relation to European cinema, see Aitken 2001.

17 See Nichols 1995, n.p.

synchronization, however, there were various projects seeking to experiment with voice, music and noise, particularly with techniques of sound collage, counterpoint and asynchrony. Some notable examples of this experimentation include Vertov's *Enthusiasm: Symphony of the Don Basin* (1931), but also GPO Film Unit productions like *Pett and Pott* (1934) and *Night Mail* (1936). Bill Nichols has noted that a second trend, from the mid-1930s, signals the establishment of what he has called »expository« documentary, characterized by direct-address style with »voice of God« commentary. This style can be observed in films like *Housing Problems* (1935) and *The River* (1937), and, after 1939, wartime propaganda films like Frank Capra's *Why We Fight* (1942-1945). Ultimately, Nichols argues, »collage, sound, and documentary became tamed, placed at the service of sponsors«.[18]

Though the history of observational and journalistic documentary has allowed for music – and sound more broadly – to be deemphasized or formalized in function, we can suggest that this historical shift was neither straightforward nor immediate. James Lastra, for instance, has usefully stressed the ongoing debates over sound perspective and continuity following the introduction of synchronized sound in Hollywood. In the struggle over 1920s reality codes, a »phonographic« model (based on fidelity to usual perception) was pitted against a »telephonic model« (foregrounding vocal clarity or intelligibility), with the latter gradually adopted as the industry-driven production convention.[19] Both models continued well into the 1930s, with technicians lobbying for a wider sound space based on multiple microphones. However, the studios' preference for sound intelligibility led to close-miked directional sound, with the microphone replicating the visual perspective. In turn, the establishment of voice-over narration in documentary could be said to be influenced by this trend towards clear, standardized vocal presentation, although documentary films generally demonstrated a greater range of accent and (un) intelligibility, particularly in post-war observational cinema.[20]

As the above has demonstrated, sound has come to play a crucial role in documentary's narration, authentication of evidence, for establishing mood and eliciting emotion. Indeed, apart from camerawork, two of documentary's most important key codes are sound-based: 1) commentary/voice and 2) authentication through sound/music.[21] Rather than further discuss recording technologies (photography, phonography, cinema), I will now focus on transmission technology. Unlike photography or cinema, which

18 Ibid.

19 Lastra 2000, 139.

20 Ruoff 1993, Chanan 2007, 116.

21 Cf. Hattendorf 1999, Corner 2002.

can capture images in time, broadcasting transmission was credited with overcoming spatial boundaries and invested with notions of liveness, intimacy and interiority. The state-administered broadcast system emerged directly out of military uses of wireless during World War I, albeit primarily for point-to-point communication for naval ships and, later, in trench warfare. The subsequent development of German broadcasting was influenced by the involvement of state postal and military authorities in telegraphy, and the close relationship between state and industry in developing communications technologies.[22] Telegraphy pioneer and postal services director Hans Bredow was placed in charge of developing a radio network from 1919. Following the establishment of the licensing system and regular programming after 1923, Bredow maintained the assertion that radio should be a public trust that would be above party politics. He also pursued a bourgeois concept of radio in terms of cultural and educational improvement (*Bildungsideal*).[23] This concept was influenced by broadcast radio's emergence at the height of inflation and Bredow joined in articulating widespread fears about the urban masses, citing radio as a domestic medium that would help to unite families and keep children away from the corrupting influences of urban streets.[24]

The institutional setting of Weimar-era radio after 1923 did not lend itself easily to creative explorations of urban soundscape, given its emergence during a tense period of social unrest, economic upheaval and political crisis. Not only was broadcasting a state-governed enterprise, but censorship boards monitored station production. This institutional context placed restrictions on the discussion of politics, and with it, the depiction of social reality and current affairs (*Aktualität*). Finally, technological difficulties with microphone sensitivity and transmission initially prevented outdoor recordings. However, the cultural production of the Weimar years is decisive for understanding the significance of sound within modern media culture and, indeed, in everyday urban life. Experimentation in broadcast content and form in Weimar Germany prefigured developments during the second half of the twentieth century, and cannot be found to the same extent in other national contexts in the same period.[25] There was a new generation of advocates for new music and radio-specific composition during the 1920s. Some

22 Cf. Friedewald 1999, 2000.

23 Hagen 2005, 75-78.

24 This institutional context involved a gradual domestication of the medium, shifting from the early male tinkers (*Bastlern*) with crystal detectors to licensed, vacuum tube sets in the domestic environment, which remained predominantly for the consumption of the middle classes in the 1920s. See Lacey 2000, 49, Marßolek 2001, 214.

25 Cf. Crook 1999, Douglas 1999/2004.

of these composers were influenced by futurist aesthetics, their own experiences of war and the sound of the modern city (usually typified by whistles, sirens, bells, engines and transport vehicles).[26] German radio is also distinct due to the engagement of writers and artists in radio production and the intellectual debates about this new medium, and their subsequent role in the development of radio as an art form, as Andreas Fickers' chapter on *Berlin Alexanderplatz* in this volume also indicates.[27] In what follows, I will investigate main currents in documentary portrayals of the urban during the decade following 1923, in terms of *documentary*, *actuality* and *authenticity*, with particular attention to the contrasting techniques of the mobile, »wandering« microphone and the »bird's eye« view.

3. Documentary Portraits of the City

The task to document the urban environment and draw on its characteristic sounds was taken up within the first months of station transmission in Germany. One of the challenges in this period was to find sufficient content to fill up broadcast schedules, which allowed for innovation and experimentation with broadcast forms, although stations drew extensively on existing cultural forms, particularly from literature and theatre traditions.[28] Here, I will focus primarily on those programmes that involved a clearly-defined project to produce a *Hörbild* (acoustic portrait) of a particular city, whether covering the entire city or a number of components that make up the whole. In this early period, such experimentation, as I will show, ranged from studio-produced portraits, modernist sound experiments and notions of documentary collage, to ambitious outdoor broadcasts that variously employed live link-ups, wax disc recording and editing techniques.

The Hamburg Norag station, which began broadcasting in March 1924, was exemplary in its attempts to foreground urban sounds in early programming. Radio-specific pro-

26 Some of these composers included Franz Schreker, Paul Hindemith, Max Butting and Kurt Weill, alongside those working in North America such as Georges Antheil and Edgar Varèse. See Hailey 1994, 14-16, Donhauser 2007, 67-125.

27 For the involvement of German intellectuals in Weimar-era radio, see Schneider 1984, Wittenbrink 2006. For a comparative assessment of German, British and American radio in this period, see Lacey 2002.

28 See, for instance, Arnheim 1933/1972, 39, 232, Cory 1994, 333-36, 346.

gramming was marked by a preoccupation with Hamburg's status as an international port and city of trade, and with it, the sounds of the harbour and maritime culture more broadly. Station director Hans Bodenstedt had previously worked as a newspaper journalist, but his concept of radio production entailed a creative rendering of the urban by means of sound. Shortly following the launch of the Hamburg station, Bodenstedt aired a radio play, *Im Hafen* (In the Harbour), which sought to draw on the »sounds of reality« (*Geräusche der Wirklichkeit*) as much as possible.[29] According to Bodenstedt's understanding, a *Hörbild* necessarily produces »a portrait that one hears« (*Ein Bild, das man hört*). It is therefore necessary that radio announcers had to learn how to create a »sonorous portrayal« of the visual (*Bild tönend malen*) using speech, music or sound effects.[30] Nonetheless, this concept of portraying the real was reliant on studio production, rather than outdoor transmission. Such conditions were observed by radio announcer and producer Alfred Braun, in a 1926 article that included sketches of the production of studio sound effects: a water basin with a fan propeller and horn signal being operated to create the departure of a ship from Hamburg harbour, a hand-operated wheel with fabric for wind, a shower onto plywood for rain, and a car horn and pressurized tank for a car.[31]

Around one year after Bodenstedt's first radio play, Norag was the first German station to conduct an outdoor transmission on 24 May 1925. The programme *Mainacht auf der Alster* (May Evening on the Alster River), was broadcast outdoors, as a way to extend the possibilities of staging radio plays against the background of the real soundscape.[32] This broadcast was enabled by cable lines connecting back to a pre-amplifier, and it was performed both for the listening audience and the large crowds that gathered around three outdoor locations. This open-air broadcast was credited with providing a sense of »three-dimensional spatial listening«, since listeners could be better encouraged to develop a fantasy of seeing the broadcast due to its spatial markings and claim on the real.[33] Norag staff member Kurt Stapelfeldt described a mixture of spoken and musical performance, set against the existing sounds of the riverside and a café terrace, and partially disrupted by wet weather conditions, which

29 Fitze 1929, 6.

30 Bodenstedt 1929, 45.

31 Braun 1926.

32 Fitze 1929, 10.

33 »*Hier draussen im Freien, wo der Schall vollkommen ausschwingt, oft auch füllend zurückkehren konnte, verhalf dies 3 dimensionale räumliche Hören der Fantasie zum plastischen Sehen*«. Stapelfeldt 1925, 1457.

required the microphone and cables to be shielded with umbrellas.[34] This programme suggests the novelty of a technological breakthrough as much as the creative investment in outdoor broadcasting of the city.

Writing several years later, Bodenstedt noted that outdoor broadcasts were still limited by the cable extension possible from pre-amplifiers (at around 500 metres).[35] He concludes with the observation that he hopes it will be possible in the future to free the microphone from a fixed location, and that short wave radio will help to allow for »significant event(s) that is instantaneously brought closer to the listener through images and commentary«.[36] Here we can already determine a modernist desire for a mobile or wandering microphone, which would allow for radio reports from cars, boats, aeroplanes and airships, in order to bring the listener to where the action was. In turn, Bodenstedt noted that the Hamburg station had conducted experiments with transmission from difficult locations – such as under water and from the air – in order to help increase the mobility of the microphone for maritime broadcasts from large ships or islands like Sylt.[37] Indeed, the introduction of mobile microphones by around 1929-1930 encouraged both the technique of a explorative or wandering microphone and that of a fixed, aerial or bird's eye view, which I will return to later in this chapter.

A strong theme in city portraits during the first years of broadcasting was the perceived need to facilitate visualization for the radio listening audience. One of the challenges, as Bodenstedt observed in 1929, was for radio reporters to learn to give their reports in visual language, and communicate their message without the benefit of headlines, layout or bold type (as with printed reportage). This notion of translating sensory impressions into verbal commentary was developed as a training technique for new radio announcers in the late 1920s. According to this *Filmfunk* (film radio) method, announcers were given the task to narrate silent film footage, in order to sufficiently hone their »radiophonic« style for live events and reportage.[38]

The emphasis on visualization, as Inge Marßolek has pointed out, was not only limited to notions of acoustic portraits but depends on a broader notion of radio as a *magic eye*. The perceived need for visualization is also evidenced by the on-going »behind the scenes« features in radio journals, which sought to familiarize listeners with station an-

34 Ibid., 1457.

35 Bodenstedt 1930, 309.

36 Ibid., 315.

37 Bodenstedt 1929, 49-50. For the preoccupation with topographies of the urban and the ongoing cultural fascination with »oceanic metaphors« developed for wireless, see Sconce 2000, 62.

38 Fitze 1929, 11.

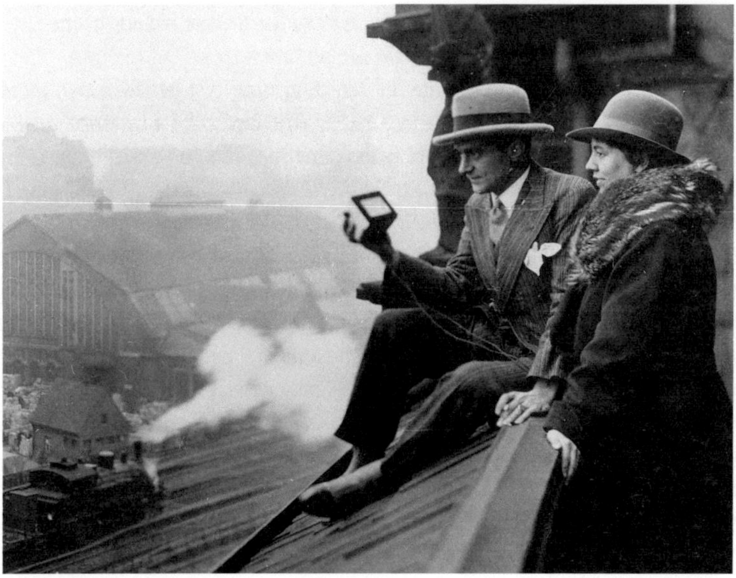

Figure 7: Ferdinand Krantze (Norag) during a reportage from the Hamburg fruit and vegetable market, 16 December 1930.

nouncers and studio production.[39] Radio journals and yearbooks also visually performed the notion of collating the sounds of the city in a montage configuration. For instance, in the 1929 Norag yearbook, an overview of the station's programming was accompanied by photographic cut-out collages of urban phenomena like trains, cars, horse races, ships, radio towers, street scenes, sailors and aesthetic configurations of (female) bodies. In the same year, the annual *Rundfunkjahrbuch* also picked up on the popular urban »symphony« theme, as discussed earlier with the city film genre. Placed within an article discussing the importance of experimentation, radio director Carl Schmidt is pictured in the foreground with a large microphone against a photographic collage depicting factory buildings, telephone lines, conveyer belts and machinery. The caption notes that radio microphones are taken to factories and shop floors in order to reveal »The symphony of work, the melody of our times« (*Die Sinfonie der Arbeit, die Melodie unserer Zeit*).[40] This emphasis on visiting various urban sites is consistent with the no-

39 Marßolek 2001, 210-212.

40 Bodenstedt 1929, 47.

tion of an explorative microphone that would unify the diverse sounds of the city with the aid of montage.

The modernist notion of visual collage in depicting the city was developed more explicitly in station programming in the late 1920s. Fritz Bischoff, for instance, drew on a concept of visuality in his experiments with sound recording on optical film celluloid. Bischoff's experimental montage *Hallo! Hier Welle Erdball* (1928) employed sound (on) film techniques to comment on radio eventness, news reportage and modern urban perception.[41] In a subsequent article, Bischoff draws on a comparison with visual cross-fading in the cinema for his notion of »acoustic dramaturgy« (*akustische Dramaturgie*), based on voice, music and sound effect.[42] Making a comparison to camera operators, the radio technician is described as turning the amplifier condenser slowly in order to fade from one sound source to another, and thereby move to the next part of the storyline. Bischoff advocates this development in terms of progress, using the controller as a technique to switch between two adjacent sound studios, as a quasi-editing device to cut between scenes.[43] As this example illustrates, Bischoff demonstrated a keen sense of the necessity for radio-specific production, and criticized the influence of the theatre in standard sound effect practices.[44] Even though the sounds of wind, machines and street noises could be used sparingly for naturalistic effect, he wrote, there was a risk that sounds that appeared realistic in reportage could seem ridiculous in a topical radio play and result in a situation that parodies reality.[45] In this context, Bischoff noted that for a recent radio play production, *Spielen aus der Zeit* (Contemporary Scenes), he and his colleagues used a switch mechanism that allowed for sound effects from their studio as well as real sounds from the street outside.[46]

A popular programme with a similar project to document the city – with a wandering microphone – was the Berlin Funk-Stunde series *Mit dem Mikro durch Berlin*

41 Prior to this broadcast, all radio plays were broadcast live and were not recorded for archiving purposes. There appear to have been two versions of Bischoff's radio play. The original version was aired in February 1928. The remaining recording fragment in the German Radio Archive (Frankfurt) appears to be derived from a second airing in November 1929. See Gilfillan 2009, 75-76.

42 Bischoff 1930, 202.

43 Ibid., 203.

44 »ein Weg von der grässlichen Theatermaschinerie, der akustischen Geräuschkulisse, den Löffelerbsen als Rieselregen, zu einer natürlichen Hörbarwendung ins akustische Spiel einzubeziehender Geräuschmittel«. Ibid., 204.

45 »...die Wirklichkeit parodiert«. Ibid., 204.

46 Ibid.

(Around Berlin with the Microphone), initiated by Alfred Braun from October 1928. This programme included reports from different (often indoor) locations as a means of taking a tour through various night-time locations. A landmark programme from Hamburg in October 1929 also conducted a four-hour reportage with almost thirty segments from different locations around the city. The *Deutscher Rundfunk* magazine reproduced the timetable for the programme, directed by Hans Bodenstedt, which preserves the notion of a potpourri of segments, each around ten minutes in length.[47] It opens with the premise that a German man in Mexico wanted to know how his *Heimatstadt* (hometown) Hamburg had become, with the first segment broadcasting his telephone conversation with the station. The reportage then presumes to travel between different sites and present the sounds of Hamburg to its listeners, who are imagined as tuning in from far away. These sounds ranged from transmissions of a chamber music concert, a variety show, sound film screening, theatre performance, short wave radio reception, fishing trawlers in the harbour, a crowded bar, and concludes with sirens at a police station and the characteristic sounds of the Michaelis church bells.[48]
One of the notions that emerged with mobile microphone techniques was that of the reporter as a kind of *flâneur*, with a microphone that could lose its way and get lost while exploring the city. This is suggested by several programmes introducing German listeners to foreign cities. The Frankfurt station Süwrag, for instance, produced a reportage from Paris, *Verirrtes Mikrophon* (Lost Microphone), in which three station reporters spoke from multiple popular locations in Paris.[49] With the more widespread adoption of wax disc recordings, in order to broadcast programmes at a later timeslot, there was more room to experiment with mobile reporting techniques. In September 1931, the Cologne radio station Werag broadcast a report by journalist Hellmut H. Hellmut as part of the series *Die Welt auf der Schallplatte* (The World on Gramophone Record). For this programme, Hellmut travelled through the streets of New York in a recording van, with a microphone used to record both ambient sound and Hellmut's description of what he saw and heard.[50] This style of recording, which seems to prefigure the subjective style of soundscape walks, was recorded onto wax records and reworked for a later Cologne broadcast.[51] In contrast to the simultaneous portrait of a city from multiple locations, the gramophone allowed for the possibility of recording a single narrative while moving

47 See the programme review: Tasiemka 1929, 1365.

48 Ibid.

49 See *Das Sechste Berliner Rundfunkjahr* 1929, 102-03.

50 Schumacher 1997, 564.

51 Cf. Järviluoma/Wagstaff 2002.

through the city, and by virtue of its recording, some possibility for alteration before airing the programme.

Norag also continued to make lyrical city portraits as late as 1931, which alternated between wandering and fixed microphone positions. One illustration of such a portrait, for which a sound recording is still available, is the approximately forty-minute programme *Städtebild Kopenhagen* from September 1931.[52] In this programme, the Norag reporter Ferdinand Krantze speaks to listeners from the centrally-located Axelborg radio building, and visits three urban sites: the shipping wharf, the town hall and central square, and the Tivoli Gardens.[53] In the programme, described in visual terms as a *Schaubericht* and *Städtebild*, Krantze begins by describing Copenhagen from his position at the central Axelborg studio. Although the programme seems to use pre-recorded segments, its stress on liveness is achieved on the basis of Krantze's delivery and the echoing characteristics of the various indoor and outdoor spaces from which the transmission takes place. At the same time, however, unlike the 1929 Hamburg programme discussed above – with its heightened sense of eventness – this programme remains conventional in its use of sounds, stilted interviews and pleas for international brotherhood, tourism and trade.

To sum up, while there were numerous projects to create urban *portraits* in early German radio, both in the studio and with location sound, the tour through the city format was considered a bit dated by 1931. Felix Stiemer, writing in *Der Deutsche Rundfunk*, suggested that these programmes initially provided the listener a chance to hear the sounds from places they had never visited, yet they had become too formulaic and undifferentiated in their portrayals. Stiemer gives the example of Alfred Braun's innovation in broadcasting from a lighthouse, but argues that this novel idea soon became a cliché when others imitated him.[54] Stiemer argues that the documentary possibilities with sound had been overrated in the first years of radio:

> One shouldn't have any illusions about this: the expressive qualities of sound have been overestimated. We know today, how even the most representative

52 See archival recording *Städtebild Kopenhagen* (8 September 1931). Deutsches Rundfunkarchiv (Frankfurt).

53 At each of the three locations, however, urban sounds are reasonably limited in scope. The programme begins with the sound of church bells and a clock chiming, and has the pretension – as part of the *Ostseejahr 1931* – of performing a link between Copenhagen and listeners of the syndicated northern German stations of Hamburg, Bremen, Kiel, Flensburg and Stettin.

54 Stiemer 1931, 31.

visual image for a city can be false; for this reason the postcard has lost its authority as the cinema has become more popular. Sound is too ambiguous and too abstract.[55]

Implicit in this argument is the suggestion that acoustic portraits of cities were not successful enough in conveying the unique essence of a city to its listener. The creative rendering of the urban had even discredited radio as a cultural form, Stiemer argues, and it was disappointing for listeners when reporters broadcast from »sightseeing« locations that were not necessarily acoustically interesting or appropriate. Despite such critiques, many commentators and practitioners defended urban portraits and reportage as a means for addressing the sounds of particular cityscapes and for accessing the social in a period of social and political crisis. In the following section, I will further investigate the concept of »actuality« in early radio practice, with particular attention to the microphone techniques used for portraying of urban social life in news and live event programming.

4. Actuality

The prevailing notion of *Aktualität* (actuality) in Weimar radio was often simplistic or naive. Indeed, one might say that Weimar radio did not really constitute a public sphere since there was still too little news and political information.[56] A clear separation of news and commentary was imposed, with centralized news service Dradag dictating the agenda and indirect style of news programming. In the first few years after 1923, politicians occasionally appeared on radio, but then only in their representative function, with their party name left unmentioned. There were few outdoor broadcasts of current affairs between 1924 and 1926 – in part due to technological restrictions – but on several occasions there were large outdoor events transmitted, such as the celebrations marking the withdrawal of French troops from the Rhineland on 31 January 1926, which featured speeches by Cologne Mayor Konrad Adenauer as well as German President von Hinden-

55 »Man soll sich keine Illusionen darüber machen: hier wird die Ausdrucksfähigkeit der akustischen Mittel weit überschätzt. Wir wissen heute, wie sogar das repräsentative Bild den individuellen Charakter einer Stadt verfälscht, weil ihm die Bewegung, das wichtigste Moment fehlt; deshalb ist auch die Ansichtskarte ausser Kurs gekommen, je populärer der Film wurde. Klang ist zu vieldeutig und zu abstrakt«. Ibid., 31.

56 Cf. Hagen 2005, 71, 79.

burg.[57] Hamburg's Norag station was pioneering in their reporting of actual events, and already had its first regular *newsreel* (*Wochenschau*) from mid-1924. But, as Bernhard Heitger has pointed out, Weimar radio was predominantly defined by an »event-related actuality, marked by an absence of politics«.[58] An illustration of this is the comment by a journalist in 1927 that radio's representation of actuality should comprise: »life, humour, unaffected naturalness«.[59] Such interpretations meant that programmes seeking out the sounds of actuality usually resulted in a visit to the zoo or circus (with the sounds of animals or exotic languages) or from public events, such as commemorations and the opening of museums and bridges.[60] One prominent feature of such actuality programming was reportage from strange locations, by taking the microphone to a mountain peak, the rooftop of a tall building or from inside an aeroplane.[61] In such cases, the microphone was usually held in a fixed position, from which the elevated reporter commented on what could be seen below from their privileged, aerial view.

The other major development – where we can observe the technique of the fixed microphone and bird's eye perspective – was in sport reporting, which became a fixture of all station programming in the 1920s. As Renate Schumacher points out, sport emerged as a major radio genre from 1925, although largely a result of its popularity with the German public, rather than its appropriateness for radio transmission. Station reporters covered rowing, and later, football, car racing, horse racing, swimming and boxing.[62] Given the difficulties in covering large outdoor events with live transmission, sport broadcasting emerged as a field of experimentation in presentation styles and technical set ups. Germany's *star announcers* such as Bernhard Ernst (Cologne), Fritz Wenzel (Breslau) and Alfred Braun (Berlin) variously placed emphasis on the sporting event, its sounds and description, and developed generic features such as unscripted commentary, interviews with athletes and multiple microphones.[63]

57 For the various withdrawal celebrations broadcast on radio, see Mohl 2009.

58 »Außerhalb der Nachrichtengebung dominierte eine ereignis-bezogene, politikfreie Aktualität«. See Heitger 2003, 399.

59 »Leben, Humor, ungeschminkte Natürlichkeit«. See Schumacher 1997, 451.

60 Ibid., 452-53.

61 Aviation, in particular, featured very prominently in outdoor reporting, particularly around 1927-1928, a period in which the Zeppelin airship developed into a national symbol. This obsession with flying is evident in radio yearbooks from the 1920s onwards. On the prominence of *airmindedness* in interwar Germany, see Fritzsche (1992).

62 Schumacher 1997, 454-59.

63 See, for instance, the editorial »In zwölfter Stunde« (1929).

It is thus in sport reporting that we can detect both the development of the wandering microphone – bringing sports listeners closer to the action – and the fixed aerial perspective from towers, providing an overview or bird's eye perspective. Of the few available recordings from outdoor sporting events, there are two instructive examples by Frankfurt reporter Paul Laven from March and May 1930. In the case of Laven's broadcast from May 1930, a national football match between Germany and England, an eight-minute archival recording gives an indication of how the specific soundscape of the Berlin stadium was presented to listeners at home. We hear cheers, applause and distant voices surrounding the reporter's microphone, which seems to be positioned above and behind the crowd. Laven's voice heightens the sense of anticipation in the match, describing what is happening prior to the game, with the arrival of players, their rendition of the national anthem and posing for photographers. On 2 March 1930, Laven's commentary for a national football match similarly provided an overall perspective for listeners at home. An archival photograph revealed his position from a tower above the crowd, from which he observed the game and crowd responses from a fixed, bird's eye position.[64]
In addition to his factual stance as a reporter, Laven's football commentary often partially adopted the stance of spectators – as the private realm – with his admissions of limited visibility, uncertainty about player names and emotive outburst following a goal:

> 11 meter line, takes a shot [...] Goal! [sound of cheers] Goal! [sound of muffled clapping][...] Attention! This is the international match between Germany and England in the Berlin stadium! All German radio stations are present. The match is in the second half. Five minutes after the start of the second half, it became 2:2 [...] Bergmeier! Shoots! Misses![65]

While Laven reported from above, looking down at both the game action and crowd

64 See archival photograph, *Paul Laven kommentiert 1930 das Fußball-Länderspiel Deutschland - Italien in Frankfurt*, 2 March 1930. Photo Archive, Deutsches Rundfunkarchiv (Frankfurt). See also archival recording, *Reportage vom Fussball-Länderspiel in Frankfurt am Main zwischen Deutschland und Italien*, 2 March 1930. Deutsches Rundfunkarchiv (Frankfurt).

65 »11 Meter Punkt, volle Schoss [...] Tor! Tor! [...] Achtung! Hier ist der Länderkampf Deutschland – England im deutschen Stadiom zu Berlin! Hier sind alle deutsche Sender. Der Kampf ist in der zweiten Halbzeit. Fünf Minuten nach Beginn der zweiten Hälfte, 2:2 [...] Bergmeier! Schoss! Aus!« See archival recording *Reportage vom Fussball-Länderkampf Deutschland gegen England im Berliner Stadion (3:3)*, 16 May 1930. Deutsches Rundfunkarchiv (Frankfurt).

below, it is clear that the commentary and atmospheric sounds heightened the immersive qualities of the game. Sports reporting, more generally, was crucial to early radio, since it

> set the precedent for a more relaxed, colloquial, and emotionally inflected form of announcing that influenced radio advertising, announcing for entertainment shows, and radio news in the 1930s and beyond.[66]

Sport reportage like this increasingly developed new strategies to heighten the representation of outdoor events, often employing sensational commentary, wandering microphones and, later, gramophone recordings and sound mixes with musical sound.[67] Such reportage gains an additional significance since its pioneers – Ernst, Wenzel, Braun, Laven – were also involved in the development of news programmes and other outdoor event reportage. Laven, for instance, began the programme *Wo uns der Schuh drückt* in 1929, which was promoted as a microphone vox pop for people from different occupations and class backgrounds, although discontinued within one year. In Breslau, for instance, current affairs and opinion were pursued with their programme *Blick in die Zeit*, which was also discontinued by 1932 following heavy criticism from conservative circles.[68] Indeed, these programmes had been initiated during the liberalization within national politics in 1928 and 1929, there was new license for politicians (particularly for social democrats) to speak on radio, along with the live broadcast of Republican commemorative events.[69] In the case of the Werag station, a new series began in 1928 that eventually appeared on weekdays under the title *Vom Tage*. This was a short and accessible presentation of current political affairs, yet it had all but disappeared in 1931 and was finally discontinued in autumn 1932. That the programme, presented by Hermann Tolle, lasted as long as it did has been since attributed to its avoidance of domestic politics and a favourable reading of economic forecasts.[70] The use of gramophone and wax disc recording for news gathering began in mid-1929. Nonetheless, this development was accompanied by concerns that recording reportage for later transmission would detach reports from the *real* and

66 Douglas 1999/2004, 202-03.

67 Schumacher, 462-66; Douglas 1999/2004, 199-218.

68 See, for instance, Heitger 2003, 400-05.

69 Marßolek 2001, 213.

70 See Heitger 2003, 401.

> that [reporters] would be released from their responsibility to present a real actuality. They would also know that they could improve their announcements at a later time.[71]

In other words, if reporters would be aware of the possibility of later re-recording, it might lead to complacency and detract from radio's purchase on liveness and eventness. Nonetheless, such programmes are consistent with those discussed in the previous section, many of which sought to unify diverse sounds of public, urban life, whether deploying a wandering microphone or notions of montage.

The gramophone news compilation appeared in earnest with Hans Flesch's end of year *Rückblick im Schallplatte* presented on 31 December 1929, a development not long after innovations in film compilations like Esfir Schub's *The Fall of the Romanov Dynasty* (1927). Flesch's radio compilation includes short clips from earlier events like Alfred Braun's forty-minute commentary from the outdoor procession and commemorative events for German Foreign Minister Gustav Stresemann in October 1929, along with clips from a car race and a ceremony laying a foundation stone in Cologne.[72] These same clips featured in several eight-minute versions of the programme presented for their novelty value to visitors at the German radio exhibition in August 1930.[73] In the year that followed, the series was aired once per month, and managed to receive permission to include short clips from Reichstag speeches.[74] What we can discern as a general trend is that most programme stations were using gramophone as part of their broader attempt to transmit from public, urban sites (*Vor-Ort-Reportage*) and thereby provide location recording of original sound sources (*Originalton*). The replaying of gramophone record mixes still proved difficult with stylus needles sometimes getting stuck during live transmission and a somewhat lengthy production process, which led to complaints about newspapers being far more up to date (*aktuell*) than such radio programming.[75]

71 »dass sie losgelöst von der Verantwortung einer wirklichen Aktualität [sprechen]. Sie [wissen], dass ihre Ansagen späterhin immer noch zu verbessern [sind]«. See *Sport auf Schallplatten. Der Deutsche Rundfunk* 1932, 9, qtd. in Favre 2007, 92.

72 See archival recording, Alfred Braun (Breslau), *Reportage von der Trauerfeier in Berlin für Reichsaussenminister Gustav Stresemann*, 6 Oktober 1929, Deutsches Rundfunkarchiv (Frankfurt).

73 See archival recording, *Rückblick auf Schallplatte* 14 August 1930, Deutsches Rundfunkarchiv (Frankfurt).

74 Schumacher 1997, 568.

75 For the various complaints on this issue during 1932, see Heitger 2003, 404.

In the context of these experiments in radio genre and style, Werag began a series in October 1929 titled *Irgendwo in Westdeutschland* (Somewhere in West Germany). The inspiration provided by montage techniques is suggested by the programme's organizing principle of the *symphony*, which should unify the multiple rhythms of industry, urban and economic life. The programme is distinctive in its attempt to foreground the sounds of work (alongside spoken interviews) as well as compositions inspired by these industrial noises. According to the script, an announcer in Cologne facilitated the switches between eleven different locations and performances in the studio, with short interludes: »That was a hammer mill, just as it revealed in reality! To artistic ears it sounds like this...«.[76] This sound bridge by the announcer introduced a large studio orchestra playing the *Hammerwerk Symphonie* (hammer mill symphony). The formal structure of the programme provides a cultural space for exchange between reportage and musical experimentation, and between the sounds of the real and their inspiration for modernist composition.

Commenting on the programme, Bernhardt Ernst noted that the introduction of amplifier vans and portable microphones allowed for more mobility during such broadcasts, and praised the decision to include five minutes from each location as heightening the enjoyment for the listener. Even though industrial recordings remain difficult, as Ernst observed, the extensive planning of the programme facilitated the idea of travel, with a wandering microphone that could transport listeners from one location to another. The notion of imaginatively taking a tour through Werag's broadcast region is reinforced by the lines marking the itinerary, with each city given a number. At the same time, however, the programme could be said to inhabit a »bird's eye« perspective, as implied by the aerial view provided by the map of the Rhineland.[77] Indeed the map depicts each city name, their geographical location and an image related to the characteristic sounds of that place.[78]

The appeal to the symphony (as a means of narrative structure) and rhythm (as a metaphor for the social) shares similarities with the city film of the same period. Instead of gesturing to industrial disputes or social-political tensions (like the city films), *Irgendwo in Westdeutschland* affirms the integrity of physical labour and regional

76 »Hammerwerk, wie die Wirklichkeit es uns zeigt! Künstlerohren hörten es so...« Ibid., 210.

77 Ibid., 208.

78 The sound sources imagined in this depiction included: cathedral clock (Münster), cattle (Westphalian farm estate), mine shaft and coke oven (Gelsenkirchen), blast furnace (Dortmund), steel pipe factory (Dortmund), forge (Essen), roller plant (Oberhausen), docklands and harbour (Duisburg), spinning and weaving mill (Viersen), main train station (Cologne) and a winery (Rhine area). See Ernst (1930).

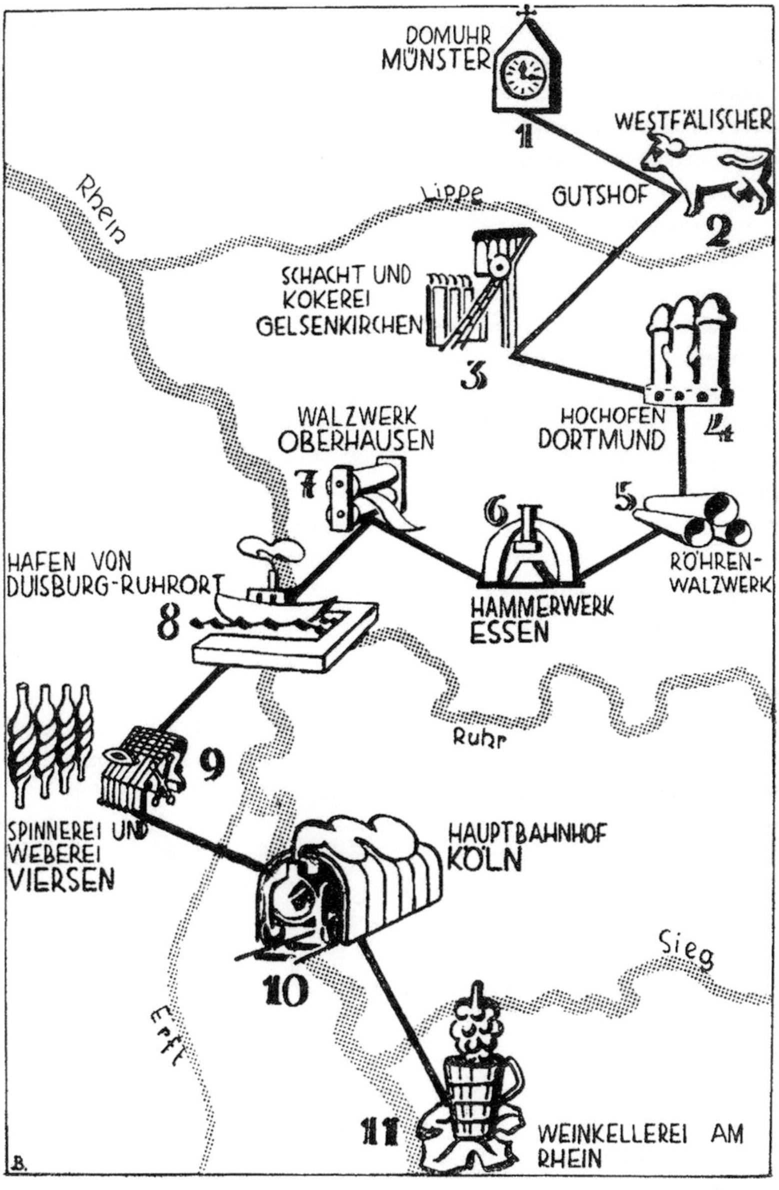

Figure 8: Image accompanying the publication Ernst, Bernhard: *Überall in Westdeutschland: Aus den aktuellen Arbeitsversuchen des Westdeutschen Rundfunks* (1930).

identity.[79] This non-confrontational position of the programme-makers was confirmed in an article by Werag reporter Bernhard Ernst, who quotes a listener letter praising the experience of listening to their region (*Heimat*) and thus being allowed to forget about political and economic crisis.[80] In this case, then, the regional appears to offer refuge from both political crisis (at a national level) and depressed urban realities (at a local level).

In sum, there were numerous experiments in event and news-related genres to heighten authenticity, eventness and »mediated liveness« in the documentary representation of urban contexts.[81] In part due to ongoing institutional and technological restrictions, the concept of actuality within radio often was quite limited in scope. With the onset of the Depression in Germany, programmes devoted to economic questions, world politics and news were under pressure from the right (particularly the National Socialist radio lobby group). Due to political fragility in 1931, state intervention became more prevalent, and state monitoring boards (*Überwachungsschüsse*) were more prone to censoring or refusing station programmes in advance.[82] In July 1932, Chancellor Franz von Papen approved extensive radio reforms that resulted in increased state ownership and unprecedented control over regional stations. These reforms meant that only government-issued reports were aired, more policing of the ban on political discussion and a clear caution on the part of station directors, who discontinued almost all innovative news and current affairs programmes during the course of 1932. While programmes like *Irgendwo in Westdeutschland* had a strong purchase on liveness and formal experimentation, such programmes increasingly focused on local sonic identities, particularly in relation to their surrounding regional areas. It is this transposition from specific urban contexts to the regional and national that I will now investigate in more detail.

79 The general address of a middle class listener is revealed by Ernst's observation that the radio interviewer had to find workers in these locations that were familiar with radio and knew what it was. This comment signals an assumption that the working classes were not necessarily considered a usual part of their imagined listening audience. Ibid., 205.

80 Ibid., 212.

81 See Auslander 1999.

82 Marßolek 2001, 213.

5. Authenticity

Representations of the city and urban life in early German radio have been framed in the previous sections with the notions of documentary and actuality. For the former, I focused primarily on city portraits, montage experiments and urban reportage techniques, and in the latter, on live event, outdoor and news reporting. This section will instead consider the significant investments in *authenticity* on the basis of voice (regional dialect) and location sounds. The Werag (Cologne) and Norag (Hamburg) stations are useful for this purpose due to their prominent but different articulations of regional identity and uses of *Heimat* and *Volk* programming. I will focus in particular on how techniques of documenting the city, with both mobile and fixed microphones, were increasingly transported to regional and national frameworks.

The establishment of the various German radio stations, from late 1923 onwards, was bound up with the idea of each station catering to a regional cultural area (*Kulturraum*). Radio, as a medium of spoken language, lent itself to programming based on local dialect (*Mundart*) and musical traditions. The West-German station was initially based in the Westphalia region, and a large portion of the station's staff and management had strong ties to local *Heimat* and *Mundart* associations. Dialect mainly featured in programmes like *Westphalian Language and Sounds* (*Westfälisch in Wort und Klang*, 1925), which were frequent but not established within a specific programme department. By contrast, the Norag station's commitment to the *Plattdeutsch* dialect was established early on, and institutionalized within a regular programme series *Heimat and Volkstum*.[83]

One major problem for the West-German station's purchase on authenticity was that it catered to two linguistic regions: Rhineland and Westphalia. Following the establishment of Cologne as the main station location in 1927, there were disagreements over the *voice of the Heimat*. The new Werag station director Ernst Hardt was half-hearted at best about regional content, and instead sought to establish station programming that was drama- and literary-oriented. Hardt, himself a member of the Social Democrat Party, subscribed to their pluralistic concept of radio, and argued that the station »would not transmit to the city or country, but would rather to the entire world«.[84] Hardt was committed to exposing listeners to art and literature, and only a fraction of the radio plays (*Sendespielen*) in 1927 were in Rhenish or Westphalian dialect.[85]

83 Karst 1984, 252-54.

84 »nicht in die Stadt oder in das Land, sondern er geht in die Welt hinaus«. Bierbach 1986, 208.

85 Karst 1984, 258.

Due to Hardt's reluctance about regional content, between 1927 and 1933, the Werag only broadcast as many dialect programmes as Norag broadcast during the course of 1927.[86] However, while dialect was perceived as too provincial, *Volk* and *Heimat* culture still formed an intrinsic part of station programming. From its inception, radio broadcasting was associated with internationalism, due to the ability to tune in to sounds from faraway places. Internationalization was experienced by many as threatening, and resulted in a renewed emphasis on German cultural forms, particularly in relation to American cultural modernity. As Adelheid von Saldern points out,

> Appeals to *Volk* and *Heimat* were seen as the only adequate response to the challenges of modernity and the so-called mass society. *Volk* and *Heimat* culture was conceived of as ›good entertainment‹.[87]

More generally, von Saldern has identified a number of similar motivations between radio station management and the promoters of *Volk* and *Heimat* culture during the Weimar era. Above all, these figures were all cultural reformers, seeking to find common ground above party politics. Many were wary of socialist-leaning worker culture, and saw *Volk* culture as a good alternative, one that was acceptable according to the dictates of state censorship. Moreover, *Volk* and *Heimat* culture appealed to bourgeois aspirations for *Bildung*, as a way of uplifting listeners' tastes.[88]

In terms of programming, the authentic sound of the *Heimat* was initially asserted in the Rhineland region with the celebratory events that surrounded the withdrawal of foreign troops from Cologne, culminating in the *Befreiungsfeier* (liberation celebration) held on 31 January 1926. It has been suggested that the gathering of large crowds outside the cathedral, the resounding church bells and presence of political leaders was given undue national significance as a syndicated mediated event and ritual performance of reoccupying the *Heimat*.[89] Prior to this high-profile event, German politicians had made efforts to preserve the *Heimat* through asserting the voice and sounds of Germans against those of French. For instance, in late 1925, Aachen politician Wilhelm Rombach (Centre Party) expressed concerns that French and Belgian stations could be heard too easily in his city. Even though Aachen was still in the occupied region, the Prussian government decided to establish a small transmission station there in April

86 Ibid., 263.

87 Saldern von 2004, 318.

88 Ibid., 319.

89 See, for instance, Eckert 1941, 23-24, 233, Göttert 1998, 426-29, Mohl 2009.

1927.[90] Despite limitations on the types of announcements made, this station provided a symbol for the presence of Germany on the local airwaves.

Dialect programming in the 1920s was varied and could take the form of radio plays, dialogues, or *potpourri* programmes. The radio play genre, although by no means standardized in its first years, was a key part of the exchange between Weimar-era broadcasting and *Heimat* culture. Radio historian Karl Karst suggests that dialect needed the radio play as a

> vehicle for *Heimat* and language preservation while for radio, dialect was the vehicle for demonstrating rootedness, folksiness and a regional connection.[91]

Potpourri broadcasts (*Mischsendungen*) often contained a mix of entertainment, fiction and factual inquiry, with segments about the local region as well as presentations, music performance and short skits. Commenting on these programmes, Adelheid von Saldern points out that variety programmes

> resulted from an attempt to satisfy disparate tastes by presenting a wide variety of material within the same program. Musical medleys containing *Volk* music and sketches relating to particular regional cultures were integral parts of these programs. They were a means of catering to the audience's diverse interests and tastes so as to bind listeners together virtually.[92]

The rise of *potpourri* and musical medley programs was less in line with broadcasters' central focus on *Bildung* than a decision to concede to popular demand for light programming and music hits.

In this period, the articulation of a social-democratic and communist idea of *Heimat*, primarily through the lens of labour, can be also discerned.[93] At Werag, for instance, social histories of the Rhenish and Westphalian region were produced, as well as worker plays, which provided fictionalized accounts about unemployment and social hardship, often with montage sequences of street and factory sounds.[94] Ernst Hardt's production

90 Bierbach 1986, 20-21, 58.

91 Karst 1984, 255, 266.

92 von Saldern 2004, 328.

93 Cebulla 2004, 169, Schumacher 1997, 221-25.

94 Examples of this genre included *Strassenmann* (Kesser, 1930) and *Der Ruf* (Kasack, 1932). See, for instance, Kreuzer 2003, 77-86.

of *Toter Mann* (1931) in particular was praised for narration of a young man who becomes unemployed, with its authenticity based on author Karl A. Düppengiesser's own experiences as a worker:

> [Düppengiesser] restricts himself to a sequence of factual events and avoids unnecessary exaggeration. The plot delivers more than just the fate of a worker, it confirms the spiritual and material need of all those whose employment is taken away from them, and who feel like a ›dead man‹ when their life's purpose is take away. […] All linguistic and acoustic possibilities are deployed in the creation of an exemplary radio play production, one that has an impressive effect. Hans Ebert has composed effective music that is based on the rhythm of work.[95]

This review suggests that the work achieves a sense of urgency, dramatizing the present-day situation of workers, by experimenting with sonic effects to, such as a symphony of factory sounds similar to the example of *Irgendwo in Westdeutschland*.[96] While we can identify such examples of programming for and about the unemployed, and interviews with workers about their situation, these programmes were produced in a time when station members like Ernst Hardt were under significant pressure. From 1930, the Nazi *Rundfunkteilnehmer* lobby group began serious agitation, as evidenced by a smear campaign in the party's *Westdeutscher Beobachter* newspaper, which attacked Hardt and several Jewish Werag employees.[97] One of the reasons Hardt was targeted was due to his efforts to commission works by modern writers such as Bertolt Brecht, and provide a platform for experimentations with radio form. Subsequently, the sounds on Werag were marked by a more cautious tone, and with the increase in censorship with new radio laws in June 1932, left-wing staff like the Marxist historian Hans Stein were rarely on air anymore. One of the other distinct changes in radio sounds in

95 »Er stellt sich ganz auf den Boden der Tatsachen und vermeidet nachteilige Übertreibungen. Die Handlung bringt mehr als das Schicksal eines Arbeitslosen, sie bringt die seelischen und materiellen Nöte aller derer, denen die Arbeit genommen wurde und die sich als 'Toter Mann' fühlen, da ihnen der Lebensinhalt genommen ist. …Alle sprachlichen und akustischen Möglichkeiten einer vorbildlichen Hörspielaufführung wurden erschöpft, so dass eine eindrucksvolle Wirkung zustande kam. Hans Ebert schrieb eine vom Rhythmus der Arbeit bestimmte, effektvolle Musik«. See *Sendung* 1931.

96 Bräutigam 2005, 381, Krug 1992, 159-90. See archival recording *Toter Mann*. Script by Karl August Düppengiesser. Produced by Ernst Hardt, Werag. 23 January 1931, Deutsches Rundfunkarchiv (Frankfurt).

97 Bernard 1997, 304.

this period is that new music and jazz featured less in programming, unless it could be disguised with the pretext of, for instance, an historical overview of popular music. The main sounds from public life remaining on radio comprised sport and *Heimat* programming like the series *Unsere Städte in Rheinland und Westfalen* and *Der westfalische Lebensraum*, both of which were launched in 1931.

To give a sense of the increased caution in programming, and renewed appeals to *Heimat*, it is instructive to revisit the programme *Welt auf der Schallplatte*, which I discussed above in relationship to urban reportage. In early 1933, an article in the *Rufer und Hörer* magazine reflected on the past 150 episodes of the programme. Even though the programme was famous for reporting the sounds of large metropolises like New York, the author emphasized that it was increasingly authentic *Heimat* sounds that the programme presented, whether in German cities or regional areas.[98] The use of the gramophone in programme production not only allowed the preservation of sonic heritage, but it heightened the ability to accrue a library of real sound effects to use later for radio plays and first-person reports, including:

> The marching of soldiers, the clopping of horses, the hissing and heaving of a railway train, the tooting of cars, the buzzing of crowds and fairground attractions, church bell tones of all kinds, waves crashing and storms.[99]

The author affirms the possibilities for recording authentic location recordings, and thus dispensing with theatre-style sound effects. However the emphasis of the article is on the ability for this radio programme to record dialect, folk song and dances, and besides German examples, the author cites recordings of regional music from Italy, France and Spain, as well as that of American Indians.

Another illustrative example of the renewed appeals to *Heimat* amongst station staff is evidence by several articles published by Fritz Worm in *Rufer und Hörer* in 1931-32. Prior to becoming the head of literature and visual art at Werag in 1931, Worm had made the series *Wert und Ehre deutscher Sprache* (1928-29) and also took part in the roundtable discussions for *Gespräche über Menschentum*. In his article on the task of *Heimat* representation, Worm takes up an idealist position, affirming the potential of radio for uniting listeners during a time of »societal, economic and worldview divi-

98 Bombe 1933, 523.

99 »das Aufmarschieren von Soldaten, das Getrappel von Pferden, das Fauchen und Stampfen einer Lokomotive, das Hupen der Kraftwagen, Volksgemurmel und Kirmesbetrieb, Glockenklänge aller Art, Meeresbrandung und Sturm« Ibid.

siveness«.[100] In this piece, Worm reframes earlier actuality experiments as being *Heimat* programming. Worm cites the example of a night-time visit to the Cologne cathedral in April 1930, in which a mobile microphone took the listener from the underground level up to the top of the bell tower. Although this programme was an experiment in reportage, Worm asserts this programme as demonstrating the intrinsic value of the *Heimat* for listeners.[101] Worm goes so far as to downplay the role of the announcer, as not required to add much extra to broadcasts of

> a celebration with dance, singing, music, the sounds of gathered audience, a funeral procession, a sport competition, a trip through a factory with the sounds of machines.[102]

More typical sounds of the Rhenish and Westphalian *Heimat* are described as including a steamship on the Rhine, songs and brass music, the pounding of machines or the goings-on of a festive crowd. While Worm vouches for a variety of sounds from the natural environment and culture (in the widest sense), his call to foreground work, the everyday, local celebrations, the voices of farmers, miners and winegrowers signals a retreat from the programmes I discussed earlier, which sought to challenge or circumvent restrictions on reporting topical news and political affairs. Radio is affirmed here as a medium that can foreground oral culture, whereby local dialect would provide comfort to listeners, if not a means for positive articulation of German cultural identity.

In the case of Norag, we can observe a similar trend with programmes like the *Das Hamburger Hafenkonzert* (Hamburg harbour concert), hosted by Karl Esmarch on Sunday mornings from 1929 onwards. This programme was remarkable given the longevity of its format over several decades, and which was based on a potpourri of concert performance, interludes, on-board reportage and interviews with seamen. At least in its first years, the programme began at seven am with the church bells of the Michaelis church, heard against music performance, usually a brass band and a festive choral, and

100 »Wie kann in einer Zeit der gesellschaftlichen, wirtschaftlichen und weltanschaulichen Zerrissenheit eine Darbietung an alle herangebracht werden?« Worm 1931, 78. Cf. Worm 1932.

101 Other examples of content described as *Heimat* programming are that of liberation celebrations in Koblenz, a Gregorian choral from Priestermund, the choir of the Cologne cathedral, a Westphalian farmers' wedding, the sounds of carnival and a procession in Paderborn.

102 »Ein Fest mit Tanz, Gesang, Musik, der Anteilnahme der versammelten Zuschauer, ein Trauerzug, ein Sportwettkampf, der Gang durch ein Fabrik mit dem Geräusch der Maschinen…« Worm 1931, 78.

harbour sounds such as a ship's whistle or horn.[103] The sonic appeal of the Hamburg harbour, as Germany's *gateway* to the world, as Esmarch suggested in 1934 was in part due to its permanent activity and romantic associations. While Esmarch emphasizes the non-political status of the programme here, it should be pointed out that the celebration of German maritime culture coincided with the expansion of Hamburg harbour activity by 1929, when it outstripped its competitors of Rotterdam and Antwerp for the first time since before 1914.[104] The radio presentation of the Hamburg harbour sounds, in this context, should be seen not only in terms of regional *Heimat* programming, but as national symbols marking the resurgence of German industry after World War I.

In this same period, the *Böttcherstrasse in Bremen* programme provides an interesting case study of an urban portrait that was permissible in this tense period of increased state intervention in radio.[105] This programme, aired by Norag on 16 June 1932, is reminiscent of some earlier *city portraits* in its premise of taking a tour through a small Bremen street. The Böttcherstrasse was a historic street, situated between the main square and the embarkment area for transatlantic ocean liners (on the New York-Berlin route). All the property had been bought and redeveloped by coffee trader Ludwig Roselius, who conceived of the street as a social-architectural blueprint for a corporatist »third way«, an alternative to the radicalized politics of the 1920s and early 1930s.[106] The radio programme in 1932 was premiered in the nine pm primetime slot, and syndicated to all German stations. This *Hörbild* was around 100 minutes in length and constructed from separate recordings on wax disc. There is a clear sense of a strict order of events, although the desired sense of liveness was disrupted by a technical default with the wax discs, which caused a sentence to be repeated during the broadcast. The use of both dramatized dialogues and limited location sounds provides a counterexample to the programmes discussed earlier, not least due to the programme's more blatant appeal to tourism, promotional culture and international trade. Despite the ban on advertising in radio, it allows for the promotion of Roselius' coffee business, as well as a projection of what the interviewer dubs a »fairytale street«. In response, Roselius

103 See archival recording *Das Hamburger Hafenkonzert*. Date unknown (circa 1929-1932), Deutsches Rundfunkarchiv (Frankfurt).

104 Kurt Esmarch, »Alle deutschen Sender übertragen Sonntags morgens das Hafenkonzert aus Hamburg«. *Funkstunde* 1934, 552, qtd. in Kiekel 2011, 14.

105 See archival recording *Die Böttcherstrasse in Bremen: Eine Strasse der Wandlungen im Mikrophon*, 16 June 1932, Deutsches Rundfunkarchiv (Frankfurt).

106 See, for instance, Henderson 1994 on the architectural reconstruction of the street and Roselius' utopian ideal, which fused the reactionary and modern.

argues that the street provides »fantasy, modernity and the homely«.[107] The broadcast's portrayal of this utopian project conveys very clearly a concept of the urban that retreats from late Weimar reality, and a reinvestment in spoken voice and dialect in terms of *authentic* German culture.

The above examples illustrate a broader trend during the last years of the Weimar era, in which political figures and the elite resorted to *Heimat* and *Volk* culture amidst cultural anxieties about national and cultural identity. The specific case of radio affirms the characterization of 1930-1932 as a transition period marked by a

> backlash against the avant-garde of Weimar and a growing role for conservative and reactionary elements in politics and culture.[108]

Journalist and satirist Kurt Tucholsky, for instance, complained that military marches and national anthems were heard more frequently on radio in this last period. On 30 January 1933, the National Socialist takeover was broadcast as a live event reportage as thousands of party supporters took to the streets of night-time Berlin.[109] The reportage commented on this procession with torchlights, which proceeded past the German Chancellery and through the triumphal archway of the Brandenburg Gate. This supposedly spontaneous performance by uniformed party members provided an important sonic assertion of the new status quo and encapsulated the contemporary slogan »Germany awakes!« (*Deutschland erwacht!*).

The subsuming of the local and regional within the national is evident in the first months of the Nazi takeover, during which the Ministry of Propaganda was established under Joseph Goebbels, followed by a National Socialist Chamber of Radio under Eugen Hadamovsky.[110] In Cologne, Werag station employees like Ernst Hardt, Fritz Worm and Hans Stein were forced out of their positions, with party member, *Heimat* culture enthusiast and archivist Dr. Heinrich Glasmeier appointed as station director.[111] On first

107 »Fantastik, Modernität und Heimlich«. The reporter visits various sites of the complex, which comprised cafes, a coffee showroom, museums, arts and crafts workshops, as well as a business club and the Institute of Physical Culture.

108 Saldern von 2004, 312.

109 See archival recording, *Reportage vom Fackelzug der SA- und der Stahlhelmformationen vor der Berliner Reichskanzlei*, 30 January 1933, Deutsches Rundfunkarchiv, Frankfurt.

110 For an elaboration on the tension between these two figures, and the eventual disbanding of Hadamovsky's organization due to Goebbels' power machinations, see Diller (1980).

111 Bernhard 1997, 305.

glance, it would appear that *Heimat* was reaffirmed, with a new station tune based on a Rhenish-Westphalian folk song, and renewed lip service played to the importance of dialect as an expression of *Blut und Boden* sentiments. However, the station came under further centralized control, and the smaller regional stations were shut down, partly because the Cologne station had more mobility with improved outdoor broadcast technology. Regional identity, according to the new political system, should not be affirmed in terms of uniqueness, but rather absorbed within the larger concept of national community.[112] A programme from 1934 titled *Stimme der Heimat* was one of the few expressions of regional identity, as rendered through an ideological lens. Taking party slogans like *Der Rundfunk geht ins Volk* as programme themes, it produced »portraits« of towns and villages that focused on artisans and local production. Dialect was sidelined in most locally-oriented programming, only featuring in songs or in catchy phrases that heightened authenticity for programmes in high German, like *Der Frohe Samstagnachmittag*.[113]

In the case of Hamburg's Norag station, we can identify two illustrative cases for how portraits of a local region were reframed within the *Stunde der Nation* (National Hour) series during 1933. This nationally-syndicated series was usually associated with simultaneity and liveness, since it was also used as a forum for broadcasting Adolf Hitler's speeches to live audiences.[114] The first programme, produced by Hans Bodenstedt, was titled *Der Harz* and involved a three-part broadcast of sounds from the Harz, Rübeland and Wendefurth areas in Lower Saxony.[115] The programme, from 4 May 1933, presents events held to commemorate international workers' day or May Day, which under the new regime was appropriated and transformed into *Der Tag der Arbeit* (National Day of Labour). Not unlike the city portrait of Copenhagen from 1931, Bodenstedt describes the natural and built environment, the towns and historical sites in this region. However, Bodenstedt's introduction also describes the actions of a *Hitler Jugend* youth group, whose spoken chorus (*Sprechchor*) and political slogans are faded in. The remaining segments alternate between commentary on National Socialism and

112 Karst 1984, 268-71.

113 See Karst 1984, 272. Karst also notes that around one-quarter of the eighty regional programmes made between 1933 and 1939 were concerned with carnival and fairs. For further discussion of carnival programming in interwar German radio, see Birdsall 2012.

114 After an unsuccessful radio recording in a studio around 1932, Hitler gave all subsequent speeches from outdoor public events (rather than enclosed studios). Cf. Hagen 2005.

115 See archival recording *Der Harz (Stunde der Nation)*, 4 May 1933, Deutsches Rundfunkarchiv (Frankfurt).

location sound recordings of folksongs and marching music, short interviews with mineworkers, foresters and railworkers, and authentic sound effects, such as woodchopping and gunshots. Bodenstedt's commentary frames the May Day events in terms of hope for Germany's future, as symbolized by the participation of the youth group in party-organized events.

Bodenstedt himself was forced to leave the Norag station in late June 1933, shortly after acting station director Karl Stapelfeldt. Shortly before his forced release, Bodenstedt participated in another *Stunde der Nation* about the island Helgoland, off the North German coastline.[116] This programme was framed more insistently in terms of the nation, opening with the *Horst-Wessel-Lied* and closing with the *Deutschlandlied* anthem, both performed live.[117] Here too, the broadcast begins with a mix of environmental sounds (birds, the wind whistling and a ship horn) and fades in a choral song about the sea. The announcer's commentary provides information about the island, interspersed with location sounds, and emphasizes the importance of this radio programme for drawing attention to the island, which had been fought over in British-German naval wars. The context of the nation is highlighted, since the German national anthem was written there in 1841, and the submarine station is a pretext for marine music performances and a speech decrying the left-wing November 1918 revolution. The local environment, in other words, is framed in terms of national commemoration of World War I losses and an *honouring of the dead* (*Totenehrung*) for SA deaths in 1923. This broadcast, too, indicates how the authentic voices and sounds of locality in event-based broadcasts were given national significance already within the first months following the Nazi takeover.

116 See archival recording *Helgoland, das Bild einer Insel im deutschen Meer,* 14 June 1933, Deutsches Rundfunkarchiv (Frankfurt).

117 On the additional of the *Horst-Wessel-Lied* to the national anthem during National Socialism, see Hermand 2002.

6. Conclusion

The above examples from the period 1923-1933 show us how *Heimat* discourse and cultural forms in radio were formulated in response to contemporary crises, and with modern techniques of reportage and gramophone prerecording. As Adelheid von Saldern points out,

> No analysis of *Volk and Heimat* culture that focuses only on dichotomies, such as modern vs. antimodern, can do justice to its multiple functions. What seemed to be symbols of durability were in fact characterized by continuing alteration.[118]

The radio programmes I have examined indicate how *Volk* and *Heimat* culture were resorted to by station managers as a safe option in a period of social tension and attacks on modern(ist) culture. As discussed here, regional identity in particular was approached differently by the Werag and Norag stations, and yet their uses of dialect and traditional culture in station programming provide insights how combinations of old and new were involved when it came to representing local environments and their inhabitants.

In sum, then, the examples discussed here illustrate the media techniques that were developed for documenting and imagining the urban during the first decade of German radio broadcasting. In particular, I have contrasted experimentation with an explorative mode of recording (with the wandering microphone) with that of an aerial perspective (or bird's eye view). Whether the microphone was presenting as wandering, reporting from overseas or venturing above and below ground, these techniques involved an appeal to listeners, sometimes venturing beyond the scope of their everyday experiences. The notions of documentary, actuality and authenticity have provided an instructive framework for examining urban representations and the frequent crossovers between outdoor sounds and studio recordings, between the visual and listening imagination. Urban sounds were foregrounded in programmes that ranged from early city portraits and dramatizations to more elaborate live transmissions from multiple sites in large metropolises like Berlin and Hamburg. The networking of the broadcast region, too, was evident in programmes like *Irgendwo in Westdeutschland* that blended industrial location sounds with musical performance in a manner similar to the *noisification* of music described in earlier chapters. The documentation and presentation of urban soundscapes, moreover, is revealed as increasingly situated within regional, national or even global frameworks. While the 1931 portrait of Copenhagen's urban sounds

118 Saldern von 2004, 319.

referred briefly to its regional context and syndication, subsequent programming – like *Die Böttcherstrasse* and *Stunde der Nation* – insistently reframed urban documentary techniques within the national. By tracing the particular investments in *Volk* and *Heimat* culture, I have demonstrated how, by 1932-1933, radiophonic depictions of the local increasingly provided a model for imagining nation through sound.

7. Sources

Aitken, Ian (2001): Physical Reality. Theories of Realism in Grierson, Kracauer, Bazin and Lukács, in: European Film Theory and Cinema. A Critical Introduction, Bloomington, 162-202.

Altman, Rick (1994): Deep-focus Sound. Citizen Kane and the Radio Aesthetic. Quarterly Journal of Film & Video 15.3, 1-33.

Altman, Rick (2004): Silent Film Sound. New York and Chichester.

Ames, Eric (2007): The Sound of Evolution, in: Maye, Harun et al. (eds.), Original/Ton: Zur Mediengeschichte des O-Tons, Konstanz, 281-315.

Arnheim, Rudolf (1933/1972): Radio. An Art of Sound, New York.

Auslander, Philip (1999): Liveness. Performance in a Mediatized Culture, London.

Barnouw, Erik (1978): The Sponsor. Notes on a Modern Potentate, New York.

Bernard, Birgit (1997): Gleichschaltung im Westdeutschen Rundfunk 1933/34, in: Breuer, Dieter / Cepl-Kaufmann, Gertrude, Moderne und Nationalsozialismus. Vorträge des interdisziplinären Arbeitskreises zur Erforschung der Moderne im Rheinland, Paderborn, 301-10.

Bierbach, Wolf (1986): Rundfunk zwischen Kommerz und Politik. Der Westdeutsche Rundfunk in der Weimarer Zeit, Frankurt am Main, Bern and New York.

Bijsterveld, Karin (2001): The Diabolical Symphony of the Mechanical Age. Technology and Symbolism of Sound in European and North American Noise Abatement Campaigns, 1900-40, in: Social Studies of Science 31.1, 37-70.

Birdsall, Carolyn (2012): Nazi Soundscapes. Sound, Technology and Urban Space in Germany, 1933-1945, Amsterdam.

Bischoff, F. W. (1930): Die Dramaturgie des Hörspiels, in: Rundfunkjahrbuch 1929, Berlin, 197-208.

Bluem, William (1965): Radio. The Forgotten Art, in: Documentary in American Television. New York.

Bodenstedt, Hans (1929): Wo eine Welle, da ist auch ein Weg, in: Norag. Das fünfte Jahr. Hamburg, 45-58.

Bodenstedt, Hans (1930): Zeppelin über Deutschland: ein aktuelles Ereignis im Rundfunk, in: Rundfunkjahrbuch 1929, Berlin, 307-15.

Bombe, Walter (1933): Die Welt auf der Schallplatte, in: Rufer und Hörer 2.11, 523-24.

Brady, Erika (1999): A Spiral Way. How the Phonograph Changed Ethnography, Jackson.

Braun, Alfred (1926): Die Aufgaben des Sendesspielleiters, in: Funkstunde. Ein Jahrbuch der Berliner Rundfunk-Sendestelle, Berlin.

Bräutigam, Thomas (2005): Hörspiel-Lexicon, Konstanz.

Cebulla, Florian (2004): Rundfunk und ländliche Gesellschaft, 1924-1945, Göttingen.

Chanan, Michael (2007): The Politics of Documentary. London.
Corner, John (2002): Sounds Real. Music and Documentary, in: Popular Music 21.3, 357-66.
Cory, Mark E. (1994): Soundplay. The Polyphonous Tradition of German Radio Art, in: Kahn, Douglas / Whitehead, Gregory (eds.), Wireless Imagination. Sound, Radio, and the Avant-Garde, Cambridge, MA and London, 331-71.
Crook, Tim (1999): Radio Drama. Theory and Practice, London.
Das Sechste Berliner Rundfunkjahr: Ein Rückblick (1.10.28-30.9.29), Berlin, 1929.
Diller, Ansgar (1980): Rundfunkpolitik im Dritten Reich, Munich.
Dolan, Josephine (2003): The Voice that cannot be Heard. Radio/broadcasting and the 'Archive', in: The Radio Journal 1.1, 63-72.
Donhauser, Peter (2007): Elektrische Klangmaschinen. Die Pionierzeit in Deutschland und Österreich, Vienna.
Douglas, Susan J. (1999/2004): Listening In. Radio and the American Imagination, Minneapolis and London.
Eckert, Gerhard (1941): Der Rundfunk als Führungsmittel, Heidelberg.
Ernst, Bernhard (1930): Überall in Westdeutschland. Aus den aktuellen Arbeitsversuchen des Westdeutschen Rundfunks, in: Rundfunkjahrbuch 1930, Berlin, 199-212.
Favre, Muriel (2007): Goebbels ›phantastische Vorstellung‹. Sinn und Zweck des O-Tons im Nationalsozialismus, in: Maye, Harun et al. (eds.), Original/Ton. Zur Medien-Geschichte des O-Tons, Konstanz, 91-100.
Fitze, Walter F. (1929): Fünf Jahre Arbeit. 1924-1928, in: Norag: Das fünfte Jahr, Hamburg, 4-12.
Friedewald, Michael (1999): Die »Tönenden Funken«. Geschichte eines frühen drahtlosen Kommunikationssystems 1905-1914, Berlin.
Friedewald, Michael (2000): The Beginnings of Radio Communication in Germany, 1897-1918, in: Journal of Radio and Audio Media, 441-63.
Fritzsche, Peter (1992): A Nation of Fliers. German Aviation and the Popular Imagination, Cambridge, MA.
Gilfillan, Daniel (2009): Pieces of Sound. German Experimental Radio, Minneapolis.
Göttert, Karl-Heinz (1998): Geschichte der Stimme, Munich.
Hagen, Wolfgang (2005): Das Radio. Zur Geschichte und Theorie des Hörfunks – Deutschland/USA, Munich.
Hailey, Christopher (1994): Rethinking Sound. Music and Radio in Weimar Germany, in: Gilliam, Bryan R. (ed.), Music and Performance during the Weimar Republic, Cambridge, 13-36.
Hattendorf, Manfred (1999): Dokumentarfilm und Authentizität. Ästhetik und Pragmatik einer Gattung, Konstanz.

Heitger, Ulrich (2003): Vom Zeitzeichen zum politischen Führungsmittel. Entwicklungstendenzen und Strukturen der Nachrichtenprogramme des Rundfunks in der Weimarer Republik, 1923-1932, Münster.

Henderson, Susan (1994): Böttcherstrasse. The Corporatist Vision of Ludwig Roselius and Bernhard Hoetger, in: The Journal of Decorative and Propaganda Arts, 164-81.

Hendy, David (2004): ›Reality Radio‹. The Documentary, in: Crisell, Andrew (ed.), More than a Music Box. Radio Cultures and Communities in a Multi-Media World, New York and Oxford, 167-88.

Hermand, Jost (2002): On the History of the ›Deutschlandlied‹, in: Applegate, Celia / Potter, Pamela (eds.): Music and German National Identity, Chicago, 251-68.

In zwölfter Stunde, in: Der Deutsche Rundfunk 7.52 (1929), 1.

Järviluoma, Helmi / Wagstaff, Gregg (2002): Soundscape Studies and Methods, Helsinki.

Kahn, Douglas (1994): Introduction. Histories of Sound Once Removed, in: Kahn, Douglas / Whitehead, Gregory (eds.), Wireless Imagination. Sound, Radio, and the Avant-Garde, Cambridge and London, 1-29.

Karst, Karl K. (1984): Regionalsprache im Massenmedium. Mundart und Dialekthörspiel, in: Först, Walter (ed.), Rundfunk in der Region. Probleme und Möglichkeiten der Regionalität, Cologne, 251-324.

Kiekel, Stefan (2011): Kurt Esmarch und das Hamburger Hafenkonzert. Über das Selbstverständnis eines Rundfunk Pioniers und die Gründe für ein erfolgreiches Format, in: Rundfunk und Geschichte, 3-17.

Kreuzer, Helmut (2003): Deutschsprachige Hörspiele 1924-33. Elf Studien zu ihrer gattungsgeschichtlichen Differenzierung, Frankfurt am Main.

Krug, Hans-Jürgen (1992): Arbeitslosenhörspiele 1930-1933, Frankfurt am Main.

Lacey, Kate (2000): From Plauderei to Propaganda. On Women's Radio in Germany 1924-1935, in: Mitchell, Caroline (ed.), Women and Radio. Airing Differences, London, 48-63.

Lacey, Kate (2002): Radio in the Great Depression. Promotional Culture, Public Service, and Propaganda, in: Hilmes, Michele / Loviglio, Jason (eds.), Radio Reader. Essays in the Cultural History of Radio, New York, 21-40.

Lange, Britta (2007): Ein Archiv von Stimmen. Kriegsgefangene unter ethnografischer Beobachtung, in: Maye, Harun et al. (eds.), Original/Ton. Zur Mediengeschichte des O-Tons, Konstanz, 317-41.

Lastra, James (2000): Sound Technology and the American Cinema. Perception, Representation, Modernity, New York.

Lessing, Theodor (1908): Der Lärm. Eine Kampfschrift gegen die Geräusche unseres Lebens, Wiesbaden.

Lessing, Theodor (1909): Ueber Psychologie des Lärms, in: Zeitschrift für Psychotherapie und medizinische Psychologie, 77-87.
Lichty, Lawrence / Bohn, Thomas W. (1974): Radio's ›March of Time‹, in: Journalism Quarterly, 458-62.
Loviglio, Jason (2002): Vox Pop. Network Radio and the Voice of the People, in: Hilmes, Michele / Loviglio, Jason (eds.), Radio Reader. Essays in the Cultural History of Radio, New York, 89-111.
Madsen, Virginia (2009): Radio and the Documentary Imagination, in: Crisell, Andrew (ed.), Radio. Critical Essays in Media and Cultural Studies. Volume I., London and New York, 207-19.
Marek, Hans-Günther (1930): Städte erstehen im Funk, in: Norag. Das Sechste Jahr. Hamburg, 73-74.
Marßolek, Inge (2001): Radio in Deutschland, 1923-1960. Zur Sozialgeschichte eines Mediums, in: Geschichte und Gesellschaft, 207-39.
McDonald, Fred A. (1979): Don't Touch That Dial. Radio Programming in American Life from 1920 to 1960, Chicago.
Mohl, Renate (2009): Wie der Rundfunk die Befreiung der Rheinlande feierte, in: Cepl-Kaufmann, Gertrude (ed.), Jahrtausendfeiern und Befreiungsfeiern im Rheinland. Zur politischen Festkultur, 1925 und 1930, 329-99.
Nichols. Bill (1995): Documentary and the Coming of Sound. URL, http://www.yidff.jp/docbox/6/box6-1-e.html (last accessed 10 June 2011).
Rosen, Philip (1993): Document and Documentary. On the Persistence of Historical Concepts, in: Renov, Michael (ed.), Theorizing Documentary. New York, 58-89.
Ruoff, Jeffrey (1993): Conventions of Sound in Documentary, in: Cinema Journal, 24-40.
Saldern, Adelheid von (2004): Volk und Heimat Culture in Radio Broadcasting during the Period of Transition from Weimar to Nazi Germany, in: The Journal of Modern History, 312-46.
Scannell, Paddy (1986): The Stuff of Radio, in: Corner, John (ed.): Documentary and the Mass Media, London, 1-26.
Schneider, Irmela (1984): Radio-Kultur in der Weimarer Republik. Eine Dokumentation, Tübingen.
Schumacher, Renate (1997): Radio als Medium und Faktor des aktuellen Geschehens, in: Leonhard, Joachim-Felix (ed.), Programmgeschichte des Hörfunks in der Weimarer Republik, Munich, 423-621.
Sconce, Jeffrey (2000): Haunted Media. Electronic Presence from Telegraphy to Television, Durham NC.
Sendung des Westdeutschen Rundfunks Köln, 23.10.1931, in: Der Deutsche Rundfunk 9.44 (1931): 68.

Simmel, Georg (1903/1997): The Metropolis and Mental Life, in: Leach, Neil (ed.), Rethinking Architecture. A Reader in Cultural Theory, London and New York, 69-79.
Stapelfeldt, Kurt (1925): Der Mainacht auf der Alster, in: Deutsche Sender, 1457-58.
Stangl, Burkhard (2000): Ethnologie im Ohr. Die Wirkungsgeschichte des Phonographen, Wien.
Stiemer, Felix (1931): Das Mikrophon wandert, in: Der Deutsche Rundfunk, 19.
Tasiemka, Hans (1929): Riesenreportage der Norag. Mit dem Mikro durch Hamburg, in: Der Deutsche Rundfunk, 1365.
Vogt, Guntram (2001): Die Stadt im Kino. Deutsche Spielfilme, 1900-2000, Marburg.
Winston, Brian (1993): The Documentary Film as Scientific Inscription, in: Renov, Michael (ed.), Theorizing Documentary, New York, 37-57.
Wittenbrink, Theresia (ed.) (2006): Schriftsteller vor dem Mikrophon. Autorenauftritte im Rundfunk der Weimarer Republik, 1924-1932: Eine Dokumentation, Berlin.
Worm, Fritz (1931): Heimatdarstellung, in: Rufer und Hörer, 76-80.
Worm, Fritz (1932): Geist und Volk, in: Rufer und Hörer, 197-202.

8. List of Films and Radio Programmes

Á Propos de Nice (1930), dir.: Jean Vigo
Berlin: Sinfonie der Großstadt (1927), dir.: Walther Ruttmann
Enthusiasm: Symphony of the Don Basin (1931), dir.: Dziga Vertov
The Fall of the Romanov Dynasty (1927), dir.: Esfir Schub.
Housing Problems (1935), prod.: British Commercial Gas Association
Man with a Movie Camera (1929), dir.: Dziga Vertov.
Moana (1926), dir.: Robert Flaherty
Nanook of the North (1922), dir.: Robert Flaherty
*Night Mai*l (1936), dir.: Harry Watt, Basil Wright
Pett and Pott (1934), dir.: Alberto Cavalcanti
Rien que les heures (1926), dir.: Alberto Cavalcanti

The River (1937), dir.: Pare Lorentz
Why We Fight (1942-1945), dir.: Frank Capra.

Die Kölner Befreiungsfeier, 31 January 1926. Werag.
Blick in die Zeit [series], from 1927 to 1932. Schlesische Funkstunde AG.

Böttcherstrasse in Bremen, 16 June 1932. Norag. Recording held at Deutsches Rundfunkarchiv (Frankfurt).
Der Frohe Samstagnachmittag [series], from November 1934 to March 1939. Werag.
Gespräche über Menschentum [series], from ca. August 1930 to 1932. Werag.
Hallo! Hier Welle Erdball, 4 February 1928. Schlesische Funkstunde AG.
Im Hafen«, 31 May 1924. Norag.
Das Hamburger Hafenkonzert [series], from 9 June 1929. Norag.
Der Harz [Stimme der Heimat series], 4 May 1933. Norag. Recording held at Deutsches Rundfunkarchiv (Frankfurt).
Helgoland, das Bild einer Insel im deutschen Meer [Stimme der Heimat series], 14 June 1933. Norag. Recording held at Deutsches Rundfunkarchiv (Frankfurt).
Irgendwo in Westdeutschland [series], from 4 October 1929. Werag.
Mainacht auf der Alster, 24 May 1925. Norag.
Mit dem Mikro durch Berlin, 27 October 1928. Funk-Stunde Berlin.
Reportage vom Fussball-Länderspiel in Frankfurt am Main zwischen Deutschland und Italien, 2 March 1930. Recording held at Deutsches Rundfunkarchiv (Frankfurt).
Rückblick im Schallplatte, 31 December 1929. Recording held at Deutsches Rundfunkarchiv (Frankfurt).
Spielen aus der Zeit, ca. 1929. Schlesische Funkstunde AG.
Städtebild Kopenhagen, 8 September 1931. Norag. Recording held at Deutsches Rundfunkarchiv (Frankfurt).
Unsere Städte in Rheinland und Westfalen [series], from 25 January 1931. Werag.
Verirrtes Mikrophon, 3 September 1929. Süwrag.
Vom Tage [series], from February 1928 to 1932. Werag.
Die Welt auf der Schallplatte [series], from May 1931. Werag.
Wert und Ehre deutscher Sprache [series], from 1928 to 1929. Werag.
Westfälisch in Wort und Klang (Westphalian Language and Sounds), 18 September 1925. Wefag.
Der westfalische Lebensraum [series], ca. 1931 to 1933. Werag.
Wo uns der Schuh drückt [series], from 1929 to 1930. Süwrag.

Radio Documentary and the Formation of Urban Aesthetics

Evi Karathanasopoulou and Andrew Crisell

1. Introduction

The *Oxford English Dictionary* defines *documentary* as »a film or television or radio programme that provides a factual report on a particular subject«. However, the etymology of the word reveals a didactic impulse that we seldom associate with modern documentary[1], for it derives from the Latin words *documentum*, »a lesson« and, ultimately, from *docere*, »to teach«. Most modern documentaries, in whatever medium, are far more tentative and oblique than we would expect *lessons* to be, and we should briefly explore this paradox.

The documentary media of film, radio and television are modern, or at least relatively recent, roughly dating from the beginning of the twentieth century. In Western Europe and the United States, where these media first developed, democracy and egalitarianism were, despite some temporary aberrations, part of a general tendency. Mass literacy spread across Europe from the middle of the eighteenth century[2] and was accompanied by the growth of mass education. As social hierarchy flattened, pedagogy, especially in respect of adult education, became less overtly didactic: in many disciplines, learning became a matter not so much of what the teacher directly imparted to the student as what the student was invited to *infer*.

1 Hendy 2004.

2 Vincent 2000.

2. Documentary Forms and Mass Media

It could be argued that when they were put to educational purposes, there was something in the very nature of the new mass media that encouraged this tendency. Film, radio and television do not readily lend themselves to didactic purposes. To deliver a lesson on the radio in the form of a speech at the microphone, or on film or television in the form of a ›*talking head*‹, seemed to place a needless curb on their pictorial or representational powers. This is not to say that such ›*lessons*‹ did not occur, but their effectiveness was soon perceived to be limited, especially when the growth of competition in the film and broadcasting industries forced their practitioners to address and exploit the things that these media could do best. Narrative attitudes became subtler, indeed overt narration was often dispensed with altogether, and in the preoccupation of the media with those facets of the physical world that they could convey so effectively, it is hardly surprising that documentaries came to be regarded primarily as factual reports. Nevertheless, the derivation of the word »documentary« is not accidental and something of its didactic impulse persists: the documentarist's aim is to tell, or perhaps more accurately show, us something new about the world. In revealing a situation, the documentarist may seek to apprise us only of its complexity: she may have no wish to shape our attitude towards it. Nevertheless, in choosing one situation rather than another, she has a rhetorical design on us of some kind, for the very choice entails a judgement – that this situation is more worthy of our attention than an infinity of other situations. And the same exercise of judgment will apply to her treatment of it: the point of view, the choice between narration or interview or dialogue, the extent to which actuality will be used, and so on.

It is a curious fact that if one types the single word »documentary« in Wikipedia, itself a reflection of popular attitudes and ideas about culture, one receives results only for the visual media. It is even more curious that while the definitions of visual documentary use terminology such as »factual«, »informative« and so on, the Wikipedia item on radio documentary associates it with drama and notions of performance.

One explanation for this may be that in the UK, for instance, governments were anxious that the BBC, a publicly funded broadcaster, should as far as possible be non-political: it was forbidden to editorialize and it was discouraged from making programmes on political themes. For this reason, early documentaries attempted few depictions of that contemporary social reality which is more often than not »political« in its significance. Moreover, documentarists like Lance Sieveking were rather more interested in exploring the nature of the new pure-sound medium than in presenting themes of a more informative and relevant nature.[3] Consequently most radio documentaries were either

3 Sieveking 1934, 15-26.

formal experiments, often of a fantastical kind, or poetic, dramatised evocations of distant events in British history.[4]

3. Radio and Urban Culture

Given the tension between the creative and didactic it is important to ask why we are interested in writing on early radio documentary in a book about the staging of urban soundscapes of the past. For the purposes of historical evidence? And if so, evidence of what – of the character of cities in the early twentieth century? Or of how the first radio producers sought to depict city life, or the way in which they approached the overall challenge of making documentaries in a sound-only medium?

We would suggest that the point of interest in these early radio documentaries lies not only in how they represented cities but in the fact that they are artefacts of high aesthetic value that were produced as *part* of an evolving urban culture. Radio has always been closely bound up with urbanity. The new twentieth-century technologies, of which radio was one, were the expressions of an increasingly industrial, and thus urban, society.

In the modern era of home-based networking and dispersed production, the links between industry and urbanization may not always be self-evident, but their historical association seems undeniable. The major traditional industries were without exception labour-intensive. Coal mining, steel making and shipbuilding took place in collieries, mills, factories and yards which accommodated large workforces, and they required the workforce to live nearby. This meant the development of urbanization, and the industrial revolution which took place between the eighteenth and twentieth centuries was accompanied by rural depopulation and an enormous growth in the size of towns and cities. Bound up with these trends was the need to improve the efficiency of communications, whether physical, such as canals, railways and roads, or intellectual, in the form of newspapers, mail and eventually forms of telecommunication including telephony and broadcasting. These abolished distance altogether, in the sense that »the same time« no longer presupposed »the same space«.[5] Cities had, of course, long been regarded as centres of economic and cultural activity. The word »urbane«, which derives from »urban«, »of the city« but which actually means »cultured, civilized«, tell-

4 Scannell 1986, 3-4.

5 Thompson 1995, 32.

ingly dates from the mid-sixteenth century. The paradigm, then, was of »city-country«, »centre-periphery«: the city as a source of commodities, wealth, and thus of political life, communicating to and nourishing those who lived beyond it. It was a paradigm that in both technological and cultural terms radio effortlessly adopted. Listeners wished to be informed of life in the cities and to immerse themselves in art forms and cultural activities that the city generated.

Even when sound broadcasting was local in range and parochial in its preoccupations, its implicit paradigm was of an urban, often metropolitan, centre radiating messages to the periphery. It is an interesting observation that in the beginning, radio was both global and local but not national.[6] When Dame Nellie Melba sang her songs into the microphone at the Marconi Wireless Works in Chelmsford in 1920, they were heard in many parts of Europe and even in America. On the other hand, when British sound broadcasting became more organized a year or two later, it was in the form of stations that transmitted to local audiences not only in London but in Birmingham, Manchester, Newcastle, Cardiff, Glasgow, Bournemouth and Aberdeen.[7] The British Broadcasting Company, which was formed in November 1922, was at first a mere organizational umbrella under which these local stations operated, but its managing director John Reith swiftly developed an integrated, national service with the metropolis, London, at its heart. By 1930 local broadcasting had vanished, while the regional services were self-evidently subordinate to the national one.[8] What Reith did, then, was to ensure that radio conformed to the prevailing cultural paradigm of city-country, except that the very nature of broadcasting – to transmit messages to *mass audiences over distances* – meant that the paradigm underwent a certain modification, a scaling-up. With radio, at least in Britain, »city-rural« becomes »metropolitan-provincial«, still of course dedicated to the diffusion of urban(e) values, but *supra*-urban in dimension. Though himself a Scotsman, Reith had little interest in local dialects, politics or cultures.

6 Hendy 2000, 21-23.

7 Hennessy 2005, 147.

8 Scannell / Cardiff 1982, 166-67.

4. Radio and Urban Aesthetics

We are therefore interested in looking at early radio documentaries as historical evidence of radio's evolution and, at the same time, at the part radio had in forming an urban aesthetics. Although the media are often seen as observers, distributors and preservers of the most recent history of humanity, the media themselves are part of a developing urbanity and their progress as part of it is documented in their content. In this age of social media, we can see with especial clarity that content is directly affected by the ways in which it is distributed. The images, the sounds, the physical shapes and the space that the paraphernalia of broadcasting and reception occupies within public and private spaces are part of this developing culture. And in recording this culture, the media are also recording *themselves*: they are documenting their own evolution. Radio sound cannot be separated from the environment in which it transmits: it is a point of view from within.

In the case of early documentaries particularly, the radio scholar cannot help but wonder whether their purpose was not simply to ›*teach*‹ people about urbanity but to flex the medium's muscles – to help producers and audience understand how radio works and to discover the true nature of the ›*radiophonic*‹. And in doing so, they were placing the medium at the centre of urban development. By becoming accustomed to radio, people perhaps came to their own understandings of what urbanity meant. Not only was the new technology coming into their houses physically and sonically, but it carried with it the expression of a new and developing aesthetic. This aesthetic derived from three different factors: the importation of unexpected sounds from other places into the listeners' familiar environments; the corporeal element, the intrusion into the home of new technological apparatus and the association of these new sounds with the apparatus; and the new possibility of montage (although at first all editing had to be done live).

British pre-war radio was disposed to celebrate not only the technology of the medium as a whole but the technologies it incorporated, especially within dramatic and documentary programming. Among the first BBC producers to explore the creative capabilities of the medium was Lance Sieveking, who was charged with the task of developing experimental forms of radio, and who, in his book *The Stuff of Radio*, meditated at length on the extent to which traditional genres and cultural forms were modified by the special technology of radio. To exploit the character of these radiogenic forms he developed what he termed the »dramatic control panel«, essentially a primitive mixing-desk which could be used to switch between a programme's various performers in different studios, and at which the producer sat like a musical maestro organizing the whole performance.[9]

9 Scannell 1986, 3-4.

Two notable productions which made use of the dramatic control panel were Charles Croker's *Speed: A Tragi-Comic Fantasy of Gods and Mortals* (1928) and Sieveking's own *Kaleidoscope: A Rhythm Representing the Life of a Man from Cradle to Grave* (1929). Described as »too purely radio to be printed for reading«, which suggests that it consisted at least as much of sound actuality and sound effects as of dialogue, *Kaleidoscope* used a cast of over a hundred in eight studios.[10] Both productions were self-conscious celebrations of the technology of radio, but the former, though describing itself as a fantasy, was evidently a reflection on contemporary forms of high-speed transport, and in that sense the celebration of another form of modern technology.[11] Such evidence as there is suggests that the audiences were more than a little baffled by these displays of technical virtuosity. In 1932, the *Radio Times* spoke of

> the enthusiastic mistakes of experiment . . . an over-indulgence in the mechanical element at the expense of the human.[12]

A very good example of how radio was used in order to understand itself is *Crisis in Spain*, a programme that was hailed as »the first British example of reportage in radio dramatic form«.[13] Produced by A.E. Harding and first broadcast on BBC radio on June 11th 1931, it illustrates a few points we shall elaborate on. For one thing, it is important to notice how the programme was described in the decade in which it was created. In a revival of it in 1938, the BBC announcer prepares us for it by saying:

> It is not so much the events themselves constituting the first Spanish revolution that are the subject of this programme, but the communication of them by means of telegraph, telephone, radio and print to a world awaiting news of *Crisis in Spain,* twelfth to the fifteenth of April, nineteen thirty one.

The programme is a multilingual montage of radio reports from around the world about the crisis in Spain. The blips of the telegraph are a constant sonic reminder of how these reports have reached the ears of the listener. The sound of trains going back and forth completes the impression of a new and fast-moving reality. Music is also blended into the miscellany, framing new sounds within an older, familiar art form. The aesthetics of a

10 Briggs 1981, 110.

11 Ibid., 120.

12 Ibid.

13 BBC announcer 1938.

new and technological era are diffused through sound waves in the form of a sonic construction that resembles not only the new sonic reality but evokes an urban architecture of intertwined train tracks, radio antennas and telephone cords. The audience is listening to the development of a new world even as it is coming out of the radio speakers. In other words, a new aesthetics of broadcasting is being created *on air* as a metaphor of the new ways in which the world is going to function. Meaning is being created not strictly in what is being said, but by means of sonic orchestration. The listeners are being given the rhythm of the new world. They are introduced to the new tempo and texture of their everyday lives. Hence the editorial decisions that are implicit in this 1931 documentary afford an insight into the creation of new aesthetic forms and understandings. Although the programme was labelled »reportage«, it is aesthetically avant-garde, replete with what we would today recognize as dramatic rather than news values. The lack of a presenter/narrator is a choice that immediately allows the listener a more inferential role and makes the narrative more fluid. As the BBC announcer very tellingly observed in 1938, the emphasis is not so much on the story itself as on how it is being told. And how the story is being told carries another story within it: a history of broadcasting and a history of the development of sonic aesthetics within a brand new urban setting. The sound here works as a narrator. The story is told *with* sound. The old form of narration is abandoned, letting the audio talk for itself. Broadcasters were moving away from the old diegetic forms and exploring the idea of the radiophonic – in this case expressed as a montage of music, sound effects and broadcast voices. This intricate new form demonstrates an excitement about exploring the medium's capabilities. The programme is *composed* and produced, using sound itself as the protagonist.

5. Conclusion. Aerial Congestion

Crisis in Spain sounds like a premonition of McLuhan's »global village«, where radio is in the centre of an increasingly better connected world. The rapid intercutting of several languages and the sounds of the telegraph demonstrate an accelerating world, and this acceleration is due to newly developed and quickly evolving communication technologies. The different languages, used interchangeably and without translation, can be heard as a metaphor of the multiculturalism that came with the development of the vast modern city.

The programme lasts for an hour, which by today's standards is long, especially for a narrator-less production. The montage of telegraph sounds, radio broadcasts and music may sound repetitious to us now, but we must make allowances for its experimental

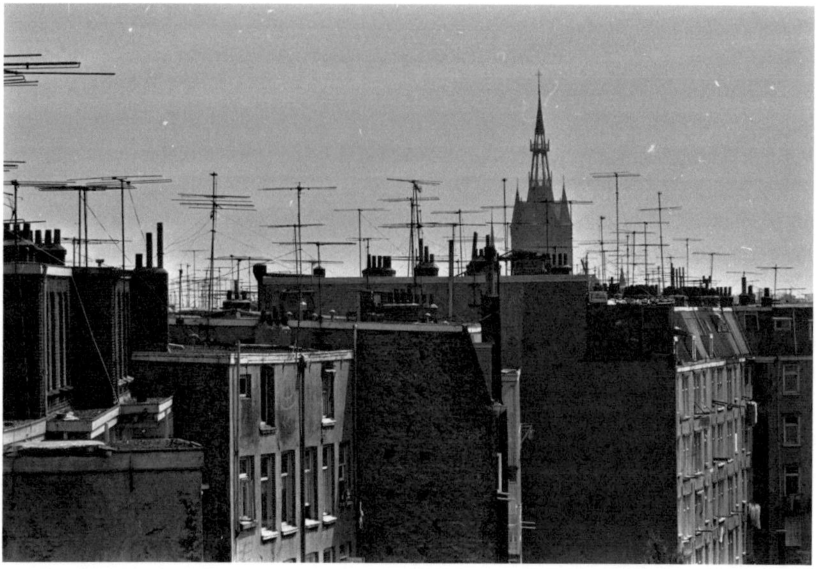

Figure 9: Antennas on Amsterdam Roofs (1964). Original caption: Today's City View:
Roofs with antennas fostering the most modern forms of leisure activities: watching television and listening to the radio.

character. It is almost as if the only reality is a *mediated* reality: without a narrator, the listener's only evidence of the events in Spain is their mediated coverage.

Radio expresses urbanity not only in terms of content but in its corporeal presence, the antennae, radio receivers, microphones, studios and so on. Aerial congestion, for instance, is very visible in the landscape of modern cities. Steven Connor has pointed out that radio interference is really a marker of the objects and bodies – including other broadcasts – that share the same space with the radio transmission.[14] Wavelength congestion mirrors all the other forms of urban congestion: buildings, people, traffic. Where you have a concentration of population, you have problems of space, a competition for space, so wavelength congestion is an extension as well as a metaphor of what happens in cities. *Crisis in Spain* expresses a congestion of voices and electronic noise that might have sounded more hectic to the early listener than to us who are used to a fast moving mediated and non-mediated polyphony. Radio thus expressed the urban in its

14 Connor 2006.

exploitation – celebration – of technology. If the links between industrialization and urbanization are on the whole evident, the relation of technology to both will be equally so. Technology made modern industry possible, and the concentrations of population that industrial production required led to urbanization. In manufacture, transport and communications, modern cities depended heavily on technology. As we argue in this essay, radio, in Britain at any rate, was an urban, indeed a metropolitan, phenomenon. Hearing the New York announcer's voice exhorting the ether with »America calling Berlin; come in London« brought to the radio some of the qualities of the telephone, and made it seem as if the announcer's voice embodied the city itself.[15]

15 Karpf 2006, 243.

6. Sources

Briggs, Susan (1981): Those Radio Times, London.

Connor, Steven (2006): Atmospherics, A lecture given at Sounding Out 3, University of Sunderland, 8 September 2006.

Hendy, David (2000): Radio in the Global Age, Cambridge.

Hendy, David (2004): Reality Radio. The Documentary, in: Crisell, Andrew (ed.), More than a Music Box. Radio Cultures and Communities in a Multi-Media World, Oxford, 167-188.

Hennessy, Brian (2005): The Emergence of Broadcasting in Britain, Lympstone.

Karpf, Anne (2006): The Human Voice. The Story of a Remarkable Talent, London.

Scannell, Paddy / Cardiff, David (1982): Serving the Nation. Public Service Broadcasting before the War, in: Waites, Bernard / Bennett, Tony/ Martin, Graham (eds.), Popular Culture Past and Present, London, 161-190.

Scannell, Paddy (1986): »The Stuff of Radio«. Developments in Radio Features and Documentaries before the War, in: Corner, J. (ed.), Documentary and the Mass Media, London, 166-67.

Sieveking, Lance (1934): The Stuff of Radio, London.

Thompson, John (1995): The Media and Modernity, Cambridge.

Vincent, David (2000): The Rise of Mass Literacy. Reading and Writing in Modern Europe, London.

http://www.thefreedictionary.com/documentary 27/12/2011.

http:/en.wikipedia.org/wiki/Radio_documentary 27/12/2011.

http:/en.wikipedia.org/wiki/Documentary film 23/01/2012.

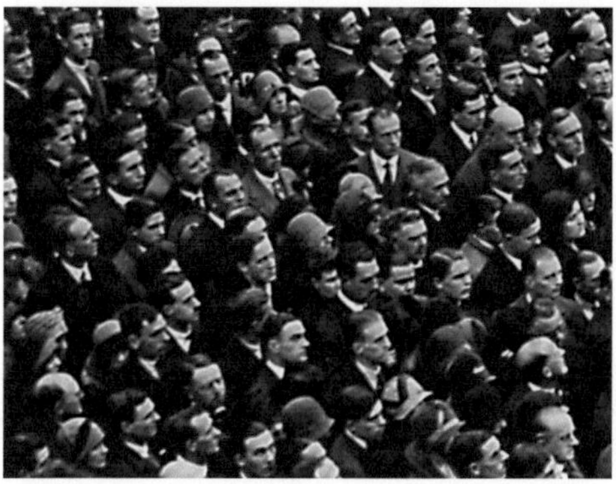

Soundscape, Landscape, Escape

Jonathan Sterne

> To hold, for example, that natural space, the space described by the geographer, existed as such and was then as some point socialized leads to the ideological posture of nostalgic regret for a space that is no longer, or else to the equally ideological view that this space is of no consequence because it is disappearing.
> (Lefebvre 1991, 190)

1. Introduction

Soundscape: no concept has proven to be more fertile or ubiquitous in the academic study of sound. The term is everywhere in sound studies, and seems somehow central to everything. For people new to the field, it provides an exciting point of entry. It also graces the covers of some of our important books. As a neologism it is immediately accessible. It sounds like what it means, even if the term lays like a blanket over a field of competing meanings.

> The term's popularity rests precisely on its ability to evoke a whole complex set of ideas, preferences, practices, scientific properties, legal frameworks, social orders, and sound that the emerging field of sound studies is – and in truth – having a difficult time getting its collective minds around.[1]

1 Kelman 2010, 228.

The word *soundscape* speaks to the physicality of sonic space; it simultaneously conveys a sense of being expansive *and* contained. Like *landscape*, it suggests spaces and people, and at once implies a point of audition and omniscience. For R. Murray Schafer, a soundscape is a sonic environment, »any acoustic field of study«[2], from physical spaces to recordings. In practice, he used *soundscape* as a total social concept to describe the field of sounds (and possibilities for sound) in a particular place, or an entire culture, »a total appreciation of the sonic environment«.[3] As Mitch Akiyama has pointed out, many of Schafer's terms exposed and inverted visual biases in the description of space:

> Landmarks become ›soundmarks‹, clairvoyance becomes ›clairaudience‹, and eyewitnesses were recast as ›earwitnesses‹. […] Schafer's neologisms alert us to the invisibility and banality of visual metaphors by reimagining language as implicitly aural.[4]

Other writers have taken up the term *soundscape* to mean many different things. Emily Thompson follows Alain Corbin in thinking more analogically, where a soundscape is

> an auditory or aural landscape. Like a landscape, a soundscape is simultaneously a physical environment and a way of perceiving that environment; it is both a world and culture constructed to make sense of that world.[5]

David W. Samuels, Louise Meintjes, Ana Maria Ochoa and Thomas Porcello consider it a useful concept because it objectifies sound for scholarly analysis. For them *soundscape*

> provides some response to the ephemerality dilemma by offering a means to materialize sounds, their interrelations, and their circulation.[6]

Outside more restricted academic definitions, the term is everywhere:

> It regularly appears in the titles of books, chapters, and articles, in the names of CDs, in the monikers of performance ensembles, in pieces by sound artists,

2 Schafer 1994, 8.

3 Ibid., 4.

4 Akiyama 2010, 56.

5 Thompson 2002, 1.

6 Samuels et al. 2010, 1.

depictions of field recordings and field recording techniques, in the vocabulary of sound design for theaters, museums or amusement parks, and even in descriptions of the work of companies that specialize in home theater installation.[7]

The term has expanded from noun to modifier and even verb: *soundscape recording* and *soundscape art* have emerged both from practices of field recording more broadly and the World Soundscape Project in particular. More recently, in a wonderful essay on Bose noise-cancelling headphones, Mack Hagood (2011) has used the term *soundscaping* to describe the process through which people shape their own sonic environments by using noise-cancelling headphones to cancel out ambient sound and substitute their own music or content, thereby asserting the privacy of their sonic space through an act of consumption. In a way, Hagood is the Yang to Schafer's Yin. Hagood presupposes the implication of scholars in modern, cosmopolitan life: they are as implicated in the desires behind noise-cancelling headphones as anyone else, carving out little, privatized spaces of quietude to keep social difference at bay and provide a space of self-constitution. His analysis is grounded in the politics of social difference around gender, race and class. *Soundscaping* is a lot of *landscaping*. For Schafer, on the other hand, *soundscape* is meant to invoke nature, and the limits and outsides of industrial society. Even as it reaches into the modern world to describe its ambiance, Schafer's *soundscape* carries with it a fairly strict – if sophisticated – antimodernist politics. For him, the concept is meant to light a way out of consumer culture. In both cases, *soundscape* is an attempt to deal with the problem of representing sonic space.

The essays in the present collection also foreground the problem of representing space. They take up the question of historical soundscapes to consider techniques of sonic-spatial representation in fiction and documentary text, film and radio. In this short meditation, I step back to consider academic traditions of representing sonic space, first in a very schematic history of the word *soundscape*, and reading it from the perspective of the so-called »spatial turn« in the humanities and social sciences.[8] In particular, I want to push beyond the widely-understood notion that soundscape is both a physical space and its representation. Rather, to follow Henri Lefebvre, I want to argue that the soundscape construct[9] simultaneously indexes a set of sonic-spatial practices, the

7 Kelman 2010, 214.

8 Gregory 1994; Warf / Arias 2009.

9 I use *construct* rather than *concept* to note that soundscape theory also has relations to particular cultural practices, and to highlight that there is no single, coherent conception of soundscape agreed upon by scholars or practitioners. Rather, like *life* to biologists, it is an intensely productive and polymorphous idea.

metadiscourses that describe them, and the cultural and sensory conditions that make it possible to – even passively – experience sonic space in certain terms.[10] Concepts of *soundscape* are artifacts of a set of professional discourses that conceive of sonic space (in this case, sound studies and acoustic ecology), but they link up with ways of perceiving and living space. In doing so, the soundscape construct moves a bit closer to the »dominant space« of our societies than we might first imagine.[11] Placed in its intellectual-historical milieu, *soundscape* is an artefact of a set of dominant ways of organizing sonic space.

2. Origins in Music and Sound Design

As a concept, *soundscape* is artefactual, which is to say it comes out of a particular cultural moment and location. In the absence of a fuller intellectual history of the term, we can turn to the *Oxford English Dictionary*, which defines *soundscape* as a »musical composition consisting of a texture of sounds« or »the sounds which form an auditory environment«. The two senses are related, and we are familiar with the latter usage derived directly from R. Murray Schafer's work, but the former sense also has a relationship to sound that is tied to place. The *Oxford English Dictionary* cites an unsigned 1968 *Time Magazine* review of Debussy's *Images pour Orchestre – Gigues, Iberia, Rondes de Printemps* as performed by l'Orchestre de la Suisse Romande and conducted by Ataulfo Argenta. The reviewer writes of Argenta:

> in this collection, he proved his mastery of the subtle colors, treacherous rhythms, and delicate contrapuntal lines that fashioned Debussy's impressionistic soundscapes.[12]

The music here is particularly poignant: the first two parts of Debussy's composition were meant to evoke his memories of England and Iberia. As a composition, at least in its conception, Debussy's triptych resonates with soundscape recordings that would be produced by acoustic ecologists in the 1970s, which also sought to document time and

10 Lefebvre 1991, 38-39.

11 Ibid., 39.

12 Anonymous review *Time Magazine* 1968, 2, cited in the »soundscape« entry in the *Oxford English Dictionary*.

place through sound. R. Murray Schafer's earliest published uses of the term *soundscape* follow this meaning. He uses the term to argue that music is not just a temporal art, but a spatial one as well: »Every piece of music is an elaborate soundscape which could be plotted in three-dimensional space«.[13]

If Debussy's musical impressionism is one origin point for our contemporary notion of soundscape, the October 4th, 1968 issue of *Time* (I have in hand the Canadian edition) gives us another. Nestled amidst stories about the U.S. presidential election, the war in Vietnam, the Black Panthers, Quebec politics, the Russian space program, and plenty of car and airline ads, one finds ads for the »Accutron: a watch that doesn't tick«, »fully transistorized« two-way radios, CBC radio's *The Sound of Sports*, the Mini-Memo portable cassette recorder, and the UNIVAC computer (»is saving a lot of people a lot of time«). Issues from other nearby weeks promote hi-fi stereo equipment. As a concept, soundscape is a creature of an orchestrated, technologized, managed sonic world.

This use of the term *soundscape* is tied to another early use of the word in radio drama. Google Books cites a 1958 issue of the BBC magazine *The Listener* which uses soundscape in a discussion of the *mise en scene* of the radio script for *The Prince of Homburg*:

> Michael Bakewell, who produced this play, was jointly responsible for the broadcast of Schiller's ›Death of Wallerstein‹ three years ago. Kleist's drama is in that tradition and Mr. Bakewell was always in command of it. His soundscape of the field of Fehrbell in presented a tremendous panorama to the mind's eye. *The Prince of Homburg* was a fine example of what the Third Programme can and should do for substantial plays that are almost unknown and underperformed in this country.[14]

Three years later, in *Samuel Beckett: A Critical Study*, Hugh Kenner used *soundscape* to describe the BBC's production and use of special effects in *All That Fall*, a Beckett play:

> Pulsating in acoustic space, the soundscape asserts a provisional reality, at every instant richly springing forth and dying.[15]

13 Schafer 1967, 13.

14 Walker, 1958, 475.

15 Kenner 1962, 171.

3. The Emergence of Soundscape as a Total Social Concept

Carlotta Darò notes two other explicit conceptualizations of *soundscape* that predate Schafer's.[16] The first is a Buckminster Fuller essay from 1966 in the *Music Educators' Journal*. Fuller borrows C.H. Waddington's concept of »epigenetic landscape« to describe the ways in which human and natural environments are co-constructed and mutually implicated.

> When, in due course, man invented words and music he altered the soundscape and the soundscape altered man. The epigenetic evolution interacting progressively between humanity and his soundscape has been profound.[17]

Fuller's use is, so far, the earliest I've found that casts *soundscape* as a total concept, analogous to a biologist's construction of *landscape*, and meant to denote the entire sonic field of humankind as it exists in dynamic relationship with nature.

In an interview with Darò, Schafer credited the soundscape concept to a series of lectures and essays by geographer Michael Southworth.[18] Reading Southworth's 1969 essay, »The Sonic Environment of Cities«, one encounters many of Schafer's core concepts and arguments in slightly different form. The essay is built around »a field study on perception of the Boston soundscape«.[19] For Southworth, noise is the main obstacle to sonic design in cities.[20] He argues that high contrast between foreground and background sounds makes sonic spaces more identifiable; and that open space and responsive spaces hold special potential for experimentation and staging sonic events. Finally, he points to sonic signs (what Schafer would call *soundmarks)* as a key to distinctive sonic experiences of the city.[21] He concludes:

> these steps toward the sonic city would not only enhance city life by helping to overcome the stress and anonymity of today's visual city, but would be one meas-

16 Darò 2012, 185.

17 Fuller 1966, 52.

18 Darò 2012, 185.

19 Southworth 1969, 49. Southworth's »field study« is remarkable for its use of disability, a topic I will consider at length elsewhere.

20 Ibid., 49, 67.

21 Ibid., 67-69.

ure for developing the sensory awareness of city residents and would provide an environment more responsive to human action and purpose.[22]

Stefan Helmreich (2011) has argued that the very idea of *soundscape* – in the Schaferian sense of »an auditory environment« owes a debt to the »stereophonic space of recorded sound«, and one can find it in the conceptual field in which *soundscape* originally operated. In *soundscape* there is a bit of a phenomenological contradiction: while the concept is designed to get people to appreciate the sounds of both natural and built environments, to confront the world *as it is*, the concept demands that the listener relate to the world as if it is a recording or composition – in short, as a *work* – but a work that is also its own means of conveyance.

Through the terms »hi-fi« and »lo-fi« Schafer explicitly conceptualizes the soundscape as a system for sound reproduction and transmission:

> A hi-fi system is one possessing a favorable signal-to-noise ratio. The hi-fi soundscape is one in which discrete sounds can be heard clearly because of the low ambient noise level.[23]

After a series of country-city and night-day comparisons, Schafer writes that

> in a lo-fi soundscape individual acoustic signals are obscured in an overdense population of sounds. The pellucid sound – a footstep in the snow, a church bell across the valley or an animal scurrying in the brush – is masked by broad-band noise. Perspective is lost. On a downtown street corner of the modern city there is no distance; there is only presence. There is cross-talk on all the channels, and in order for the most ordinary sounds to be heard they have to be increasingly amplified.[24]

Following Sophie Arkette, Ari Y. Kelman has suggested this aesthetic is tied to an «urban prejudice» in Schafer, a fundamental hostility to the way cities sound.[25] This is not necessarily the case, since Southworth's critique of urban noise is not based in an anti-urban bias. While Southworth presupposes the modern city (even has he critiques it), Schafer

22 Ibid., 70.

23 Schafer 1994, 43.

24 Ibid.

25 Kelman 2010, 217; Arkette 2004.

connects his sense of urban alienation with a preference for smaller social groupings, as when he argues that the human voice is the ideal »module for acoustic design«.[26] Hi-fi and low-fi also invoke a way of listening that had come into vogue as Schafer's ideas first came together, and the terms suggest another important branch in a critical genealogy of *soundscape*. As Keir Keightley and Tim Anderson have argued, hi-fi systems – and the ideas attending them – have their own place in the post-war cultural complex. Writing of the specifically American scene, they describe the ideals of hi-fi as intimately connected both with the escapist affects of middle class masculine domesticity, and in a critique of mass culture.

> The conception of home audio as a masculine technology that permits a virtual escape from domestic space is a significant development in the history of sound recording. Before World War II, the phonograph and recorded music were not especially associated with men. By the 1960s, however home audio sound reproduction equipment had hardened into masculinist technologies *par excellence*.[27]

Periodicals of the time trumpeted the hi-fi boom as a rejection of the mass, feminized tastes embodied by television; in these articles,

> high fidelity is cast as high, masculine, individualistic art, and television is portrayed as low, feminine, mass entertainment.[28]

Magazines and advertisements presented hi-fi as cultivated, sophisticated and edifying. A hi-fi system was said to promise access to the extremes of experience and an escape from the world of middlebrow taste and the levelling effects of mass culture. It offered opportunities for immersion and transcendence through contemplative listening. Although the hi-fi would eventually be superseded by the stereo system, the same logics of gender, domesticity and escape operated within the widespread commercial discourses accompanying stereo equipment.[29] And although Schafer's politics are clearly both anti-modernist and anti-consumerist, he makes use of the same language of escape. The very definition of the hi-fi soundscape borrows its morphology from the aesthetics of the hi-fi record and hi-fi system in the bourgeois living room.

26 Schafer 1994, 215-16; see also Sterne 2003, 242.

27 Keightley 1996, 150; Anderson 2006.

28 Keightley 1996, 156.

29 Anderson 2006.

Eric Barry locates early spectacles of high fidelity sound reproduction in the longer history of the American technological sublime (see Marx 2000 for the classic discussion of this phenomenon). Like railroads and electric lights, hi-fi audio systems became »objects of aesthetic pleasure and symbols of American identity«.[30] Although Schafer is Canadian, this same logic of technological sublimity in pastoral space guides the move to soundscape recording. Despite somewhat different political motivations, early World Soundscape Project recordings took up on the rhetoric already present in recordings like Emory Cook's *Rail Dynamics* (ca. 1951), which was meant to document both locomotives that were going out of use, and the spectacle of hi-fi sound reproduction. Thus, hi-fi culture informed both the theory and practice of work in acoustic ecology, at least in its earliest formal statements and documents.

Schafer also directly links *soundscape* to avant-garde trends in twentieth-century composition and the practices of music appreciation most often connected with Western art music. He writes:

> the opening out of the space-time containers we call compositions and concert halls to allow the introduction of a whole new world of sounds outside them.

From John Cage's *4'33"* to Pierre Schafer's *musique concrete* to electronic and tape music, Schafer finds inspiration from the canon of 20th-century experimental composers.[31] Further, we should

> regard the soundscape of the world as a huge musical composition, unfolding around us ceaselessly. We are simultaneously its performers, its audience, its composers. [...] Only a total appreciation of the acoustic environment can give us the resources for improving the orchestration of the soundscape.[32]

30 Barry 2010, 116.

31 Schafer 1994, 5.

32 Ibid., 205-06. Schafer is ambivalent on the position of the composer. Although he clearly identifies with it in his writings and musical work, he is also aware of the position's limitations, noting that acoustic design should »never become design control from above«, and that acoustic designers must understand »acoustics, psychology, sociology, music and a great deal more besides« – a demand that still too often goes unfulfilled in real world practices of acoustic design.

While Schafer's desire to exceed the »spatial frame« of the concert hall might be echoed by a critic of the western tradition like Christopher Small[33], Schafer retains an essentially heroic model of the composer, »separate from both performer on the one hand and audience from the other«[34], dressed now in the clothes of the acoustic designer. Sounds are

> the natural raw material of the art, are thought of as mere recalcitrant matter, to be put in order by the force of will and intelligence.[35]

In this way, Schafer follows from the Cagean tradition. As Cage wrote in 1937,

> The sound of a truck at fifty miles per hour. Static between the stations. Rain. We want to capture and control these sounds, to use them not as sound effects but as music.[36]

4. Conclusion

The sociogenesis of *soundscape* is typical of twentieth-century sonic concepts. Everything psychoacoustics claims to know about hearing in the state of nature was the result of interactions between ears and telephones.[37] So too, *soundscape* was shaped by a relationship to recording, reproduction and western art music concert tradition. The desire for aesthetic of purity that animates Schafer's cultural criticism seems entirely of a piece with talk of high fidelity and stereophonic reproduction of concert music in the 1950s and 1960s, as well as the cultures surrounding institutionalized art music at the time. *Soundscape* implies a way of listening to compositions – a rapt, total attention

33 Small 1977, 25. This desire to escape the frame may well have met an equal, opposite force in Schafer's conception of soundscape recording as putting a frame around sounds (Akiyama 2010, 57). In turn this opens out into the spatial implications of the enframing function of sound technologies more generally (Heidegger 1977; Sofia 2000; Sterne 2012).

34 Schafer 1994, 29.

35 Ibid., 30. This stands in contrast to Barry Truax's more interactional model, where »the individual listener in a soundscape is not engaged in a passive type of energy reception but rather is part of a dynamic system of information exchange« (Truax 1984, 11).

36 Reprinted in Cage 1961, 3.

37 Sterne 2012.

– and a sense of the world that is much like a compositional work. A soundscape is a totality, whether we consider that totality something small, like a recording, or something huge, like the entire sonic airspace of a town, country or culture.

Of course, nobody has to accept Schafer's definitions or ways of thinking about soundscape. Looking around today, we have precisely the opposite situation: the term seems to have almost infinite plasticity, and indeed many writers reject some part of Schafer's terminology or politics, but still find the term incredibly useful. The term is everywhere capable of being mobilized to support a host of positions regarding sonic culture. The term's almost instinctive appeal – to academic writers, journalists, acoustic ecologists, architects, composers, musicians, music critics, software designers, students and many others – has more than a little connection to the habitus that subtends the hi-fi systems and concert halls that Schafer explicitly invokes in his work. In its stretch toward totality, the term reaches out toward omniscience, but like all proposals for transcendence, it can at best offer a situated transcendence.

5. Sources

Akiyama, Mitchell (2010): Transparent Listening. Soundscape Composition's Objects of Study, in: Revue D'art Canadienne / Canadian Art Review 35.1 (July 14), 54-62.

Anderson, Tim (2006): Making Easy Listening. Material Culture and Postwar American Recording, Minneapolis.

Arkette, Sophie (2004): Sounds Like City, in: Theory, Culture & Society 21.1 (February 1), 159-68. doi: 10.1177/0263276404040486.

Barry, Eric (2010): High-Fidelity Sound as Spectacle and Sublime, 1950-1961, in: Suisman, David / Strasser, Susan (eds.), Sound in the Age of Mechanical Reproduction, Philadelphia, 115-38.

Cage, John (1961): Silence. Lectures and Writings, Wesleyan University Press.

Darò, Carlotta (2012): Avant-gardes Sonores En Architecture, Paris.

Fuller, R. Buckminster (1966): The Music of the New Life, in: Music Educators Journal 52.6-7, 52-68.

Gregory, Derek (1994): Geographical Imaginations, Cambridge.

Hagood, Mack (2011): Quiet Comfort. Noise, Otherness, and the Mobile Production of Personal Space, in: American Quarterly 63.3 (September 15), 573-89.

Heidegger, Martin (1977): The Question Concerning Technology and Other Essays, New York.

Helmreich, Stefan (2011): Submarine Media. Conducting Undersea Ethnography (with Notes on Underwater Music), presented at the History and Philosophy of Science Seminar Series, November 16, McGill University.

Keightley, Keir (1996): »Turn It Down!« She Shrieked. Gender, Domestic Space, and High Fidelity, 1948-59, in: Popular Music 15.2, 149-77.

Kelman, Ari Y. (2010): Rethinking the Soundscape. A Critical Genealogy of a Key Term in Sound Studies, in: The Senses and Society 5.2, 212-34. doi:10.2752/174589210X12668381452845.

Kenner, Hugh (1962): Samuel Beckett: a Critical Study, London.

Lefebvre, Henri (1991): The Production of Space, Cambridge.

Marx, Leo (2000): The Machine in the Garden. Technology and the Pastoral Ideal in America, Oxford.

Samuels, David W. / Meintjes, Louise / Ochoa, Ana Maria / Porcello, Thomas (2010): Soundscapes. Toward a Sounded Anthropology, in: Annual Review of Anthropology 39 (October 21), 329-45. doi:10.1146/annurev-anthro-022510-132230.

Schafer, R. Murray (1994): The Soundscape. Our Sonic Environment and the Tuning of the World, Rochester Vermont.

Small, Christopher (1977): Music-Society-Education, London.

Sofia, Zoe (2000): Container Technologies, in: Hypatia 15.2, 181-219.
Southworth, Michael (1969): The Sonic Environment of Cities, in: Environment and Behavior 1.1, 49-70.
Sterne, Jonathan (2003): The Audible Past. Cultural Origins of Sound Reproduction, Durham.
Sterne, Jonathan (2012): MP3. The Meaning of a Format, Durham.
Thompson, Emily (2002): The Soundscape of Modernity. Architectural Acoustics and the Culture of Listening in America 1900-1930, Cambridge.
Truax, Barry (1984): Acoustic Communication, Norwood.
Walker, Roy (1958): Drama. Prince in Darkness, in: The Listener 59 (1511), 473-75.
Warf, Barney / Arias, Santa (eds.) (2009): The Spatial Turn. Interdisciplinary Perspectives, New York.

The Corporeality of Listening

Experiencing Soundscapes on Audio Guides

Holger Schulze

1. Introduction

In the early summer of 2012, I intended to visit a museum on the history of Berlin that would offer its visitors audio guides, one of the most significant practices of staging sound as cultural heritage today. I started with a quick search, contacted several relevant institutions and tried to gain more detailed information on whether and how they used audio footage in their audio guides. After actually having visited several of these museums, however, I felt somewhat disappointed. Neither the public, state-run museums such as the *Märkisches Museum, Gedenkstätte Berliner Mauer* and *Museum Haus am Checkpoint Charlie,* nor the private, commercial exhibitions such as *The Story of Berlin* or *Historiale Berlin* happened to stage the sounds of Berlin's past in a detailed, contextualized manner. The audio guides provided fairly traditional narrations and explanations, while the more innovative museums offered so-called *Hörbänke* (listening benches) at which one can select sound files from a computer workstation, or 3D-/cinematographic arrangements. These 3D presentations seemingly aimed at overwhelming the visitors with symphonic soundtracks and a few very brief historical sound bites, touching them in a Hollywood-movie kind of way. The Berlin history museums and exhibitions thus offered either recordings of historical urban sound without much narration, or narration without much urban sound.

The field of producing and using audio guides is developing fast, however, both in technological and cultural terms. In recent years, many museums and exhibition venues have exchanged their previous communication equipment and systems for audio guides. Ever since we have smartphones in our pockets, small companies have started developing and selling applications or *apps* that can guide individual listeners through an exhibition. For the upcoming years we may even expect more radical changes in the hardware and software for audio guides. The use of individual devices with internet access and audio players will challenge the production companies of audio guides to adjust their products to these new technologies and cultural practices: be it new generations of smartphones or tablet computers, be it a pair of glasses with tiny monitors and internet access (e.g. Google's *Project Glass*) or devices we cannot yet imagine at this moment. Other novel technologies may open up options for transmitting exhibition sound files directly and without headphones or similar tools to the ears of the visitors.

This contribution, however, will not focus on the latest gadget or on the most recent attempts to capture and mediatize the sounds of the urban past, but aims to reflect on rather fundamental issues concerning the experience of audio guides. First of all, listening to sounds is – from the perspective of cultural anthropology[1] – dependent on both the physical environment in which it takes place and the state and habitus of the person listening. This means that sound should not only be studied in terms of its physics, as has been the tradition of Hermann von Helmholtz and subsequent scientists, but also in terms of its *corporealization*: sound manifests itself via individual listening bodies and material entities. Sounding and listening are grounded in the corporeality of human beings.

This holds even more for an era in which media technologies are the predominant means of creating sound events: they are able to transform any physical space (for instance by THX-Sound, 5.1 or 7.1 Surround Sound or Waveform Synthesis) into a theatre optimized for specific forms of sound reproduction, sound transmission and culturally highly specific listening situations. Yet these technological advancements, as may be surprising to some, only highlight the importance of the individual body and sensory experience. This phenomenon touches upon an often recurring dialectic in the cultural history of technology and media: the more advanced the technological means are in transcending the body's supposed limitations, the more the users of such novel technologies will become aware of the specific capacities and incapacities of their body and the subtlety of their corporeal sensibility.

This is also true for the field of sound reproduction. As the German musicologist, media studies scholar and jazz musician Rolf Großmann has argued in a seminal article on the notion of »performance practice« (»*Aufführungspraxis*«), musicology and sound studies urgently need to reflect on the material conditions of the listening experience in its widest sense. He proposes to use the term »auditory dispositive« (»*auditives Dispositiv*«) for this goal, critically drawing on work from film and media studies.[2] The notion of

1 Schulze 2012. For a general introduction to the research strategies and questions of cultural and historical anthropology, see Wulff 2009.

2 Großmann 2008.

dispositive has quite a long history by now. In 1970, film studies scholar Jean-Louis Baudry published an article on the *dispositive* of cinema.[3] This *apparatus*-theory analyzed the film reproduction and projection technology, the cinema setting and the viewing position of the film audience as key elements of a material and situated disposition, a *dispositif* (in French) for experiencing film. Subsequently, Michel Foucault[4] unfolded the wider historical potential for analyzing material setups in various social and cultural environments; and Giorgio Agamben[5] interpreted the dispositive as a highly important thought style for a material cultural history, one being close to the theory of *affordances*.[6] Ever since Rolf Großmann's article was published in 2008, there has been an ongoing discussion on how to apply the dispositive to musical and auditory phenomena.
In this short contribution I will explore the experiential side of contemporary auditory dispositives for listening to audio guides. Doing research in the field of a cultural anthropology of sound, I aim to shed more light on the individual experience of listening whilst using new audio technologies. I will try to evaluate what aspects of individual listening experiences are dominant and need to be taken into account while conceptualizing and designing audio guides with soundscapes – soundscapes of the past or on other topics. What happens with a listener when he or she listens to recordings in the specific techno-cultural circumstances of using audio guides in museums and exhibitions? While Michel Chion's notion of »point of audition« (»le point d'écoute«[7]) presupposes a rather static position for those listening to sound productions, we really need to reflect on *mobility* when analyzing audio guides, as these were already mobile devices before the advent of digital technology, podcasts and online streaming services. To analyze this particular situation of listening, I distinguish three *auditory dispositives* that are relevant for listening to recorded and post-produced soundscapes in exhibition spaces:

(a) the spatial dispositive
(b) the temporal dispositive
(c) the narrative dispositive

In doing so I will draw on an earlier study by Hanna Buhl and me on conceptualizing, producing, implementing and using audio guides in art museums and art galleries,

3 Baudry 1970.

4 Foucault 1978, 1980.

5 Agamben 2009.

6 Gibson 1977.

7 Chion 1991, 91.

for which we interviewed several producers, designers and marketing staff from audio guide companies such as *Acoustiguide* and *Antenna Audio* as well as curators and artists.[8] In addition, my reflections on the audio guide listening situation are informed by participant observation and interviews with a small number of visitors in exhibition venues. Moreover, I critically observed myself, with a phenomenological twist, as an auditory-oriented visitor. How did I experience the use of audio guides?

2. The Spatial Dispositive. Listening in Architectural Bodies

As soon as we decide to put on the headphones of an audio guide device, we transform the auditory dispositive in which we will listen to what is offered to us for the next few minutes or longer. This decision is quite obvious. But it comes with another decision – one that is often ignored. We aim to leave the experience of the architectural space of the exhibition space behind and concentrate on the experience of a sonically constructed space.

It depends on the type of headphones connected with the audio guide whether or not it allows us to hear the environmental noises of the exhibition environment in addition to the recorded soundscapes. Often, however, we receive a pair of closed headphones that provide an intense and concentrated auditory experience. This approach expresses a certain negation or even rejection of architectural space: the original spatial dispositive seems to be subjected to the technologically constructed auditory dispositive. But is it really possible or even desirable to fully separate these two aspects of the listening experience? How might designers of audio guides take the combined experience into account? We might even add another layer to the complexity of the listening situation. We should not only reflect on the consequences of listening in architectural space when tuning into the sonically constructed space of the audio guide, but also flag the spatial dimensions of the sonic construction itself. When listening to an audio guide, we attune ourselves to a mixed media constellation of (a) materially fixed recorded objects, their reverberation time in recorded space and its specific reflections, and (b) of a technical apparatus with its specific means of sound transmission and reproduction. This two-fold spatial auditory dispositive is made out of software, hardware and *buildware*. This is embedded in the third layer of the auditory dispositive: the venue's architectural space in which we

8 Buhl / Schulze 2012: a study executed in order to develop an instruction manual for museum and exhibition audio guides, commissioned by the German *Stiftung Zuhören*.

move around. Such proprioceptive aspects of listening are a key interest in the cultural anthropology of listening.

To understand the aural aspects of the spatial dispositive we have just outlined, Barry Blesser and Linda-Ruth Salter's detailed exploration of the aurality of space is highly informative. In their volume *Spaces speak, Are you listening? Experiencing Aural Architecture* (2007), Blesser and Salter succeed in trying to find a specific terminology and adequate *Denkfiguren* to describe and to analyze what can be aurally experienced in architectural space. Rather than relating to the long research tradition of building or room acoustics, they undertake the daunting effort to combine highly diverse approaches to listening from anthropology, acoustics, psychoacoustics and sound art aesthetics with the aim to enlighten the experiential side of hearing in architectural spaces.

Blesser and Salter have introduced key terms and concepts for our discourse on spatial listening. They for instance propose to differentiate between the *acoustic horizon* and the *aural arena*[9] in which we hear in any given situation. While the first concept describes the horizon to which our ability of hearing stretches out, that is to the most distant point audible, the second notion refers to the ideal area in which we may enjoy a sound source to its full extent. These concepts help us to distinguish the desired situation of audibility in exhibition venues from the non-intended noises and sounds entering this aural arena from our acoustic horizon. When designing soundscapes for audio guides, knowing the acoustic horizon of the listener at the most important spots in the exhibition is thus highly important.

Next to the acoustic horizon and aural arena, Blesser and Salter also have an ear for how an architectural artefact is *sonically illuminated*[10] by sounds in everyday listening situations: a building does not sound as such, but needs to be activated via sound. So in designing and furnishing an architectural space for a specific use, such as a museum or a gallery, it is important to take into account how the materials and constructions used will sound when sonically illuminated by the activities of visitors: their footsteps, talks and screams in the aural arena. Even if this seems self-evident, museum spaces have only rarely been designed for their auditory use – it is the spectacle most of them have been built for. Yet when their visitors enter the exhibition space with an audio guide, this past still speaks and should be considered when designing audio guides and redesigning museum spaces. It is such a historicity of listening that is another key interest of the cultural anthropology of listening.

As said, the situation gets even more complicated when we start thinking about how

9 Blesser/Salter 2006, 20-34 passim.

10 Blesser/Salter 2006, 12-19 passim.

we listen to a recorded audio file of a soundscape – of an urban past for instance – in the architectural space of the venue: this creates overlapping acoustic horizons as you hear the physically present soundscapes of the venue and the geographically and, perhaps, temporally remote post-produced soundscapes of another place. This layering forces the listener to find his own mode of listening in this auditory dispositive of technology and architecture. And as the architectural body of the venue is often muted and replaced by the sound recording's space, through playing the recording at full volume, even non-hearing the venue may be an issue and affect people's ability to orient themselves in a particular space. To our surprise, however, our interviews with experts showed that spatial layering effects in using audio guides are rarely reflected on by designers and curators, even though our participant observations clarified that the effects created irritating or even downright disturbing experiences for the visitors.[11]
We may understand slightly better now how this layering and refocusing of sound sources works. One layer is very present as it is almost directly projected onto our eardrum through technology, while the other is acoustically rather distant but bodily very present through our acoustic horizon and orientation on spatial reflections. This intriguing layering of recorded and life sounds elicits a form of *re-spacing* via the auditory. The term *spacing* has been being introduced by the sociologist Martina Löw in her *Raumsoziologie* or *sociology of space*. In her work she explains how humans generate the physical experience of a space via a process she calls *spacing*, a process of assigning the location of objects, human beings and activities, and via something she calls *Syntheseleistung* or *achievement of synthesis*: forming a coherent impression of a situation in space.[12] When listening to recorded soundscapes, mediated via the sound quality of the audio guide device, and within the soundscape of an exhibition venue, we experience spacing and re-spacing: the listening body keeps recalibrating itself.

2. The Temporal Dispositive. Listening whilst Moving and Our Experience of Time

As soon as we begin walking through in an exhibition space with an audio guide, we transform our auditory experience: we are not standing still or sitting quietly in a chair, but are constantly moving with our whole body, keeping our balance, finding our way

11 Buhl / Schulze 2012.

12 Löw 2001, 158f.

through the exhibit and keeping in touch with friends, colleagues or family members accompanying us to the venue. And while doing so, we do not only move in space, but also in *time*: we structure the temporal experience of the exhibition. In case the audio-guide does not impose a particular temporal order, for instance when we have downloaded a collection of audio files on a mobile device, we may even choose to listen prior or after visiting the actual exhibition. We may check the files and skip through them to get an overview of the audio guide. This is another aspect of the audio guide rarely reflected upon by designers and curators these days.[13]

In our exploration of a given space via our individually moving bodies, we create a characteristic, often very personal temporal structure – we may opt for a *flanerie*, we may wander through the exhibition, or cross it while in haste. Each visitor will find her or his own temporal order in response to previous experiences in exhibition venues, idiosyncratic ways of walking and moving through a space, and expectations about the current situation.[14] But how can we unravel such temporal qualities of listening in the specific *aural arena* and the wider *acoustic horizon* of exhibition venues? Where can we find inspiration to do this?

Taking the epistemological value of *artistic research*[15] seriously, we can learn from existing discourse about the auditory art of the *soundwalk*. Established by the composer Hildegard Westerkamp in the 1970s and recently revived by the highly immersive and site-specific *Audio Walks* and *Video Walks* of Janet Cardiff[16], the soundwalk is still the most extensively explored and the most intensely discussed entrance into the phenomenon of listening and into the auditory exploration of a particular space while moving in time. In 1974, Westerkamp described how a soundwalk may start – by focusing on the sounding body:

> Start by listening to the sounds of your body while moving. They are closest to you and establish the first dialogue between you and the environment. If you can hear even the quietest of these sounds you are moving through an environment which is scaled on human proportions. In other words, with your voice or your footsteps for instance, you are ›talking‹ to your environment which then in turn responds by giving your sounds a specific acoustic quality.[17]

13 Buhl / Schulze 2012.

14 Buhl / Schulze 2012.

15 Shiner 2001, Borgdorff 2007, Schulze 2005.

16 Cardiff 1991.

17 Westerkamp 2007, 49.

This suggestive recipe for a soundwalk shows how the foundation of listening whilst moving can be found in the bodily, corporeal qualities of listening. In surprising accordance with today's performative theories of the body, Westerkamp advises us to listen to what our body has to say. In the words of the philosopher Jean-Luc Nancy:

> Being a body is being a certain tone, a certain tension. I'd also even say that a tension is also a tending.[18]

Listening whilst moving begins with listening to our body moving; and as we are moving our concentration wanders, digresses and re-focuses according to the speed, the rhythm, and the environment of our movement. We explore, or should explore, space by attuning our body to the continuity and discontinuity, the rhythm and pitch of environmental sound:

> Go out and listen. Choose an acoustic environment which in your opinion sets a good base for your environmental compositions. In the same way in which architects acquaint themselves with the landscape into which they want to integrate the shape of a house, so we must get to know the main characteristics of the soundscape into which we want to immerse our own sounds. What kinds of rhythms does it contain, what kinds of pitches, how many continuous sounds, how many and what kinds of discrete sounds, etc.[19]

Human beings, and this is another insight coming from the cultural anthropology of sound, form a resonating continuum in relation to their environment, including its animals, objects and technologies – a continuum which can be explored, affirmed, neglected and ignored. In making audio guides, however, it may be more interesting to seek connections or dialogue with the rhythm of the body moving through the exhibition than to simply ignore it.

3. The Narrative Dispositive. Listening to Stories

The voice in our head, or better: the voice that gets transmitted into our earlobes and onto our eardrums via vibrating membranes, this voice tells us a story, documentary

18 Nancy 2008, 134.
19 Westerkamp 2007, 49.

or fictional, when we listen to an audio guide. He or she narrates a certain vocalized thread about each exhibit, be it an artwork, a document or something else: an auditory narrative.

But our concentration on oral narrations is not stable or linear; it shifts and rises, it gets weakened and strengthened by signals and signs in our environment and by our own proprioceptive dynamics during the day. We might digress, we may have associations with the work we need to do in the next hours or days, or we are reminded of friends, lovers, family or colleagues, or imagine or personal fictitious worlds. Somehow we have to integrate and synthesize the sounds of the venue, the voice on our audio guide, the story it tells and our shifting imaginations. Again, we may look for a source of inspiration in how sound artists deal with it by returning to composer Hildegard Westerkamp:

> So far you have isolated sounds from each other in your listening and gotten to know them as individual entities. But each one of them is part of a bigger environmental composition. Therefore reassemble them all and listen to them as if to a piece of music played by many different instruments. Do you like what you hear? Pick out the sounds you like the most and create the ideal soundscape in the context of your present surroundings. What would be its main characteristics? Is it just an idealistic dream or could it be made a reality?[20]

After all, soundscape compositions also presents narrations or stories, even though these are quite different from traditional stories of the theatrical and plot-based radio play. It is fictitious in a way that also a large archive of experimental radio play in the tradition of German radio is, bearing the proud name of *Hörspiel*, literally »playing with listening«. The element of play, the ludic character of the experimental *Hörspiel* is so central to its auditory aesthetics that its fiction element cannot be reduced to a plot line. The most prominent examples of this tradition rather tend to be examples of sound poetry and audio art: from Ernst Jandl's *Das Röcheln der Mona Lisa* (1970) to John Cage's *Roaratorio* (1979), to Heiner Goebbels' *Wolokolamsker Chausee* (1989), Andreas Ammer's *Radio Inferno* (1993) and Michaela Melián's *Memory Loops* (2010). Recorded sound has a strong and prolific agency in all of these sound pieces, and audio guide designers might use sounds as actors in similar ways.

The most advanced approach to such an aesthetic of *auditory agency* has been proposed by DJ and music critic Kodwo Eshun in his volume *More Brilliant Than The Sun.*[21]

20 Westerkamp 2007, 49.

21 Eshun 1998.

In this inspiring and provocative work, Eshun tells about his auditory journeys into the realms of free jazz, of advanced electronic dance music, and of imaginary and real new technologies, drawing connections with the history of electronic warfare, the invention of new musical instruments, and even extraterrestrial creatures. Eshun presents no orderly, scholarly writing, but an unconventional and rhapsodic account of his individual, yet not arbitrary, listening experiences, in stream-of-consciousness style.
The auditory agency Kodwo Eshun wants to celebrate is probably best explained with the following quote about the character of overwhelming auditory experiences, and how traditional acoustic theory fails to capture what sound does to him when DJing:

> There is no distance with volume, you're swallowed up by sound. [...] Not only is it the literary that's useless, all traditional theory is pointless. All that works is the sonic plus the machine that you're building. So you can bring back any of these particular theoretical tools if you like, but they better work. And the way you can test it out is to actually play the records.[22]

In its highly affective intensity, sound is to Eshun a truly corporeal and kinaesthetic agency that should be listened to but needs no superimposed narrative. Today's music is itself already »more conceptual than at any point this century«, full of »thought probes waiting to be activated«.[23] The soundscapes of artists are often of that kind, and museum curators might find inspiration in their sonic fiction.

4. Conclusion. Designing Audio Guides with the Corporeality of Listening in Mind

We may actually design more interesting audio guides if we take the spatial dispositive, temporal dispositive and narrative dispositive of the listening experience in exhibition venues more seriously. It would not only have consequences for the audio guide recordings, but also for the venue space itself. It would be highly worthwhile to design both the exhibition and the audio guide recordings in such a way that acoustic horizon and aural arena interact in fruitful ways – that people are still able to orient themselves in the museum but also to focus on the audio guide. We should, in other

22 Eshun 1998a, 188f.

23 Eshun 1998a, -003.

words, sonically illuminate the museum in such a way that not only its spaces speak, but also its audio guides – and convincingly so. We may also learn from artists who create soundwalks. If they are used to tune into the sounds of their environment, they also invite us to connect the tempo and rhythm of our walking body to the tempo and rhythm of the environment. Sound artists, finally, may teach audio guide designers to listen to sounds as narratives without words, as fictions affecting our body without presenting an argument.

5. Sources

Agamben, Giorgio (2009): What Is An Apparatus. And Other Essays. Palo Alto.

Baudry, Jean-Louis (1970): Cinéma: effets idéologiques produits par l'appareil de base, in: Cinéthique 7-8, Paris, 1-8.

Borgdorff, Henk (2007): The Mode of Knowledge Production in Artistic Research, in: Gehm, Sabine / Husemann, Pirkko / von Wilcke, Katharina (eds.), Knowledge in Motion. Perspectives of Artistic and Scientific Research in Dance, Bielefeld, 73-80.

Blesser, Barry / Salter, Linda-Ruth (2006): Spaces Speak – Are You Listening? Experiencing Aural Architecture. Cambridge/Mass.

Buhl, Hanna / Schulze, Holger (2012): Eine Ausstellung hören? Über Notwendigkeit und Qualität von Audioguides, in: Stiftung Zuhören (ed.): Audioguides (working title), Munich (in preparation).

Bull, Michael (2007): Sound Moves. iPod Culture and Urban Experience. London.

Cage, John (1979): Roaratorio. Ein irischer Circus über Finnegans Wake, Cologne / Stuttgart.

Cardiff, Janet (1991ff.): Walks. (online: http://www.cardiffmiller.com/artworks/walks)

Chion, Michel (1991): Audio-Vision. Sound on Screen, New York.

Einstürzende Neubauten (2000): Die Befindlichkeit des Landes, in: idem, Silence is sexy. London, Track 8.
(online: http://www.neubauten.org/pages/lyrics/silenceissexy/track_008.html)

Eshun, Kodwo (1998): More Brilliant Than The Sun. Adventures in Sonic Fiction. London.

Foucault, Michel (1976): Histoire de la Sexualité 1. La Volonté de Savoir. Paris

Foucault, Michel (1980): Power / Knowledge. Selected Interviews and Other Writings 1972-1977. Edited by Colin Gordon. New York.

Gibson, James Jerome (1977): The Theory of Affordances, in: Perceiving, Acting, and Knowing, Hillsdale NJ.

Gibson, James Jerome (1979): The Ecological Approach to Visual Perception, Boston.

Goebbels, Heiner (1989): Wolokolamsker Chausee I-V, Baden-Baden / Munich / Cologne.

Gopinath, Sumanth / Stanyek, Jason (2012): Oxford Handbook of Mobile Music Studies, Oxford.

Großmann, Rolf (2008): Verschlafener Medienwandel, in: Positionen – Beiträge zur Neuen Musik (2008), H. 74 – Dispositiv(e), 6-9.

Hartmann, Maren / Rössler, Philipp / Höflich, Joachim R. (2008): After the Mobile Phone? Social Changes and the Development of Mobile Communication, Berlin.

Jandl, Ernst (1970): Das Röcheln der Mona Lisa, Munich / Frankfurt am Main / Hamburg.

Kassabian, Anahid (2002): Ubiquitous Listening, in: Hesmondalgh, David / Negus, Keith (eds.), Popular Music Studies. London, 131–142.

Kassabian, Anahid (2012): The Soundtracks of Our Lives. Ubiquitous Musics and Distributed Subjectivities, Berkeley CA.

Löw, Martina (2001): Raumsoziologie, Frankfurt am Main.

Melián, Michaela (2010): Memory Loops, Munich.

Nancy, Jean-Luc (2008): Corpus. Translated by Richard A. Rand. New York.

Schulze, Holger (2005): Was sind Sound Studies? Vorstellung einer neuen und zugleich alten Disziplin, in: Vorkoeper, Ute (ed.): Hybride Dialoge. Rückschau auf die Modellversuche zur künstlerischen Ausbildung an Hochschulen im BLK-Programm »Kulturelle Bildung im Medienzeitalter«, Bonn, 79-83.

Schulze, Holger (2012): The Sonic Persona. An Anthropology of Sound, in: Michels, Axel / Wulff, Christoph (2012), Exploring the Senses, London / New York / New Delhi (in preparation).

Schulze, Holger (2013): Adventures In Sonic Fictions, in: Journal of Sonic Studies, Leiden / Nijmegen (in preparation).

Schulze, Holger / Wulff, Christoph (2007). Klanganthropologie. Performativität – Imagination – Narration, Berlin.

Shiner, Larry (2001): The Invention of Art – A Cultural History, Chicago.

Siegert, Hubertus (2001): Berlin Babylon. Berlin.

Westerkamp, Hildegard (1974/2007): Soundwalking. Originally published in Sound Heritage, Vol. III, No. 4, Victoria BC, 1974 (revised 2001 & published in: Carlyle, Angus (ed.), Autumn Leaves, Sound and the Environment in Artistic Practice, Double Entendre, Paris, 2007, 49, online: http://www.sfu.ca/~westerka/writings%20 page/articles%20pages/soundwalking.html)

Wulff, Christoph (2009): Anthropologie. Geschichte, Kultur, Philosophie, Cologne.

The Eleventh of the Eleventh of the Eleventh

The Theatre of Memorial Silence

Ross Brown

1. A Reverberant Scenography

> […] here is one of the great paradoxes, that no broadcast is more impressive than the silence following the last dashing strokes of Big Ben. Its impressiveness is intensified by the fact that the silence is not a dead silence, for Big Ben strikes the hour, and then the bickering of sparrows, the crisp rustle of falling leaves, the creasing of pigeon wings as they take flight, uneasy at the strange hush, contrast with the traffic din of London some minutes before. Naturally, vigilant control of the microphone is essential. Audible distress near to the microphone would create a picture out of perspective as regards the crowd's solemn impassivity and feelings. Our job is to reduce all local noises to the right proportions, so that the silence may be heard for what it really is, a solvent which destroys personality and gives us leave to be great and universal.[1]

So writes an unnamed BBC radio technician in the Radio Times, November 1935. He is describing the British broadcasting institution that is the annual outside broadcast of the two-minute silence from Whitehall, in London. The silence is the centre-piece of a service of remembrance commemorating the moment of the end of World War I at 11am on the 11th November 1918: the eleventh hour of the eleventh day of the eleventh month.[2] The memorial one or two-minute silence[3] stages urban sound through a some-

1 Cf. anon. in Gregory 2001.

2 See The Royal Channel 2010.

3 Henceforth referred to as the *silence*, to indicate reference to the memorial event.

what different interaction of media and participatory audience to the other examples in this volume. This chapter takes this quotation as a starting point for considering this interaction as a ritual practiced in an intermedially immersive theatre.

Silences, since Princess Diana and 9/11, have become increasingly popular ways of simultaneously celebrating community and paying respect to the memory of the dead. One might argue that they are part of a humanist meta-narrative; emblems of hope – of a human capacity, after all is said and done, to be solemn, great and universal in times that might otherwise suggest humans to be decadent, violent and trivial. On the other hand, one might argue that they are rhetorical rallying devices, both affective, in that they are *stirring*, and effective, in that they harness and focus crowd behaviour away from unruliness. There has been a proliferation of *silences* as preludes to large sporting occasions. Broadcast commentators, rather than relate their comments to any specific theme of remembrance, tend routinely to comment on how the »impeccable« observation of a *silence* is testament to the »spirit of the game« or to a human bond that transcends partisanship. It is as if remembrance is a McGuffin[4], in screenwriting terms, and the real point has more to do with being, presence and community.

There is a feeling of power in being part of a noisy crowd, and an electricity – a potential power – in being part of a resolutely silent one. We use the term *silent*, not *still*, or *mute* and this choice of word presupposes this power to be in some way aural. Belgian dramatist Maeterlinck, a pioneer-modernist of the dramaturgy of sound, suggested crowd silence produces a *dread*, suggesting the *sublime* – a term more usually associated with the thunderously sonic.[5] Reporting on the first Armistice Day event, the Guardian describes a »silence that was almost a pain«.[6] What interests me here, is the way in which some of this affect, which one might associate with the subjectivity of *presence*, permeates the airwaves and survives mediatization. The Armistice Day/Remembrance Day[7] service in Whitehall is as much a mediatized, and acousmatically dislocated event as it is a *live* one. The first ceremony in 1919 was filmed and John Reith placed microphones on the scene in 1929.

4 »McGuffin, n. In a film (now also in a novel or other form of narrative fiction): a particular event, object, factor, etc., initially presented as being of great significance to the story, but often having little actual importance for the plot as it develops.« Oxford English Dictionary 2001.

5 »We can bear, when need must be, the silence of ourselves, that of isolation: but the silence of many – silence multiplied – and above all the silence of a crowd – these are supernatural burdens, whose inexplicable weight brings dread to the mightiest soul«. See Maeterlinck 1897, 7-8.

6 Schwartz 2011, 612.

7 Up until World War II, the service was held on Armistice Day (11th Nov), but after WWII on the nearest Sunday – Remembrance Sunday. The format of the ceremony remains essentially the same.

The known presence of outside-broadcast microphones at a public event can in itself add electricity and a sense of supernatural *reach*. Indeed, the Whitehall event is definitively intermedial. Like Big Ben heralding the New Year, it is an icon that provides the focal point for simultaneously observed *silences* around the country, thus uniting them in a single networked theatre. One can imagine the extended radio/TV audience as part of the event's crowd, and their networked silence as part of its silence. In aural terms, the space of the event is an intermedial hybrid, and *presence* becomes an intermedial subjectivity.

In the quotation above, the broadcast is described as having an intensive impressiveness, as though some of the overpowering effect of being part of the crowd silence permeates the airwaves. The *silence*, is »not a dead silence« and neither are radio broadcasts of it anything like the radio anathema of »*dead* air«.[8] The silence is clearly not that of malfunction or missed cue, but is planned and staged; ceremonially framed. The whole ceremony works towards the stroke of eleven. In the run up, the ear is prepared by an aural choreography of brass band music; the shouts, boots and rifle noises of military drill; the close-miked, almost liturgical, vocal cadence and rhythm adopted by outside-broadcast commentators on State occasions. Big Ben chimes the hour, then on the first strike of eleven a huge and distant artillery salvo sounds. The chimes in total take just over one minute, during which the crowd stands silent. Then the two minutes of silence themselves. Then, to mark the end, another artillery salvo followed by the *Last Post* on bugles. For three minutes in total the crowd has withheld movement and voice and tried to constrain coughs and sneezes. The centrepiece of the show is a performed cessation of human activity, but the silence it produces somehow consists of exhilaratingly *live* air. This is captured by live microphones situated in lively acoustic surroundings. In its aurality, this service for the dead, for all its solemnity, is a celebration of liveness, both in acoustic and broadcast terms. Liveness is in the sparrows; leaves; pigeon wings taking flight; the humanity of the unavoidable cough; the dropped object. It is also in incidents. Hillel Schwarz describes down-and-out veterans shouting »anyone want a medal« and Communists singing working class anthems during Great Depression *silences*.[9] In the 1969 BBC broadcast, protests may be heard.[10]

Ordinarily in ritual practices, the accidental has little value.[11] The *Radio Times*' philoso-

8 On television, of course, the moving pictures continue through the silence and the soundtrack, while still powerful, is not as intensively impressive as the radio broadcast.

9 Ibid.

10 These may be heard in the archival recording of the 1969 silence on Semper, 2001i.

11 E.g. Jackson 1968, 293: »noises of the natural world cannot, because of their unpredictability, be of much value in the ritual itself, which is not to say they are without significance; they may be ›commanded‹ to appear in the ritual«.

pher-technician, however, relates such noises to the *silence's* power to transform participants into a liminoid state where personality is dissolved and they become »great and universal« beings. He notes that noises »intensify« the silence, and that it is an important part of his work to depict them in relation to what he calls the »solemn impassivity« of the *silence*. In audio terms, we are told, there is a »right« proportion and the noises must be kept quiet, but in comparison to what? The sound of a crowd actively being solemn? A crowd enacting silence? How are such things represented in sound?

Sonic artist Jonty Semper's *Kenotaphion* project[12] helps answer these questions by abstracting memorial *silence* from the subjectivity of its time/space-specific context. Its second 2001 release, a double CD, comprises nothing more than a chronological compilation of archival recordings of the Whitehall silence, from 1929 to 2000, each cropped from the beginning of Big Ben's chimes to the end of the silence. Set out this way, as concrete art objects, one is able to listen to them analytically without being caught up in the liveness of the event and one hears that mediatized silence is *not* silent in two ways. As well as the background noise and occasional, faint accident in the acoustic scene represented in the recordings, there is also the noise inherent to the broadcast and archival media. This decreases over the time-period represented on the discs, whereas the background noise of London increases. Typically, in a signal-to-noise ratio, the *signal* is a foregrounded sonic event-object such as speech that stands out discernibly as a distinctive figure against the noise or interference of the medium. In a recording of silence the presupposed foregrounded signal is a programmatic *absence* of any foreground event or figure – an open channel without voice. On the face of it, all there is is noise: noise of the medium and somewhere amongst it, the little noises that, in this case, point to the unheard presence of the silent performance.

In the earlier recordings archived on phonographic disc, little random incidents such as birds or coughs or other indeterminate noises are only distinguishable from the crackle and hiss because they bear, faintly trailing behind them, the tell-tale reverberation of materially spatial events – specifically spatial events in a built environment of stone and glass. Some of the little pops and clicks reverberate and others do not, and the skilled psychoacoustic brain seizes on this differential. The dry clicks are assigned to the noise category, and the reverberant noises are used, like acoustic light, to paint an impressionistic mental image of a London scene. These reverberations are loaded with meta-data. They say that this is not just any sonic snapshot of London, but specifically a silenced scene in London. Their acoustic signature is clearly that of a built environment.

12 Cf. Semper 2001i and 2001ii.

Yet the fact that they are audible within the same earshot as the soundmark of Big Ben implies that this is not an everyday, central London soundscape, but one unusually stilled. This, the skilled, listening brain subconsciously deduces, is an area one would normally expect to be so loud in the daytime that one would not be able to hear sounds such as birds or coughs, reverberating discretely between stone and glass facades and the tarmac surfaces. Where there are large crowds present, as there were in the earlier Whitehall *silences*, it perhaps also recognizes the acoustically-absorptive presence of massed human bodies. In the later recordings, the permadrone of London traffic is noticeable, but the general city noise sounds strangely distanced. The spatially vast decay of the artillery blast on the first strike of eleven, which rolls around the sky almost like thunder, describes the greater London. But once it has subsided and the chimes end it is as though a perimeter has been set up around the vicinity of Big Ben and the noise pushed back to allow small noises to assume unfamiliar significance.

This then is how the silenced crowd is represented in sound: through the reverberations its stillness reveals. The »right« aesthetic proportion, to which the radio technician refers, is represented, in audio-production terms, through privileging the reverb – the *wet*, indirect wave-arrivals of accidental noises – over the *dry*, direct wave-arrivals of the noises themselves. The silence that destroys personality and effects universality is thus found in tiny sonic accidents reverberating more prominently than they should.

2. A Ritual Aural Practice

The sonic art performance project *Noise Memory Gesture: the Theatre in a Minute's Silence* (or NMG)[13] considered the *silence* as a form of theatre in which the participant is both performer and audience. Working with Butoh and Laban techniques[14] and vari-

13 Brown 2006.

14 Two different dance approaches to regimenting the body in relation to external space and sensory environment. Butoh reaches into psychic interiority and uses intense sensory imagination to embody the sensory environment in the nervous system. Laban is more consciously analytic, and concerned with spatial relationships and the effort of interaction between body and environment. Laban identifies goals when moving in relation to the environment, whereas Butoh abandons them. The relationship between these techniques allowed us to explore the dialectic in the *silence* between the aural and the performing body. See also Brown 2009.

ous practice-based workshop methods, it researched the *corporeal performance* of the silence. In particular it examined familiar postures of remembrance: standing still, hands clasped in front of the body; head bowed slightly, eyes focused indeterminately on the middle distance. It found these to be imitative of a known genre, learned through the media, and as a corporeal regime, inseparable from the notion of *observance* of silence as a performance of remembrance. In other words, participation in the silence was an act of the *whole-body* and this active corporeal attitude was somehow inseparable from the *aurality* both of mediatized and in situ *silences*. As part of its methodology, NMG convened *flashmob silences* in a range of noisy urban environments, including extremely noisy places such as Piccadilly Circus where the traffic and tourist chatter carried on and there was no measurable drop in sound level during the *silence*. Participants nevertheless reported that actively standing still in an adopted posture seemed to contribute to an increased panoramic awareness of surrounding acoustic space and a sense of more highly-defined sonic detail. The act of performing a *silence* through purposefully being still, while bearing in mind a given memorial purpose, seemed to transform a place even of oppressive, intrusive and sometimes painful noise-levels into a high-definition, surround-sound experience. Some reported this as pleasurable; others, in some way *fitting* to the memorial purpose.

This interdependent combination of corporeal performance, focused psychological purpose (remembrance) and heightened awareness of the circumstantial materiality of audition seems to fit with what Holger Schultze, *pace* Rolf Grossman, theorizes elsewhere in this volume, as an auditory *dispositive*. Schultze deals with three kinds of dispositive (spatial, temporal and narrative), each of which might be said to describe a theatricalization of the soundscape through the adoption of an apparatus that estranges one's subjectivity from its usual relationship to the sonic environment. I think the memorial silence is representative of a fourth kind of auditory dispositive: that of a ritual aural practice. Its apparatus is the active participation of its subject as performer enacting, with body and mind, a role prescribed by generic convention (i.e. standing in memorial silence, in this case). We saw before that in order to represent the *silence* on the radio, the philosophizing radio technician knew he had to represent correctly a ratio between circumstantial noises and what he called a »solemn impassivity«. This is the *ritual dispositive*: a dialectic between one's own enactment of a role and material circumstance. In assuming this dispositive by ritually observing a *silence*, one does not merely experience the incidental, insignificant, inconsequential or passing sonic environment within a picture frame, as one might in a sound walk or at a performance of John Cage's *4'33"*. Rather, one's auditory experience (and the event's audience) is subsumed *by* the performance and becomes part and parcel of a whole system of enactment. A preordained time and duration; a generically prescribed way of standing; a given topic to be remembered; the required assuming of a sombre mood: all of this comes together in a performative silence that is the agent of liminoid transformation the philosopher-technician calls a »solvent«.

Aside from any feelings of greatness and universality, this effects a hyper-real auditory experience of accentuated panorama and heightened definition, but it does more than this. Traffic and other noises cease to be incidental or background to a moment of everyday transience and instead become monumental (if one can imagine a monument as an immersive rather than an objective figure). Like the Cenotaph *empty tomb*, it is an anonymous monument; a cipher for multifarious losses; a collectively-pondered dialectic between a ritually performed organization of stillness in the living and the noises of a world carrying on. Indeed, participants in the *silences* staged by the NMG project tended to report that they were too engaged in the material »here and now« to experience thoughts that one might describe as memories. If anything, some said, the performance silenced memories.

3. The Dramaturgy of the Silence

Lévi-Strauss observed that silence receives meaning only by contrast to noise.[15] As Hillel Schwartz describes, the first time the Armistice Day silence was observed, it was set up and announced not only by canon blasts and clock chimes, but by all manner of noise making, including »loud guns, rockets, churchbells, fire alarms, and a bugler on the balcony of Selfridge's Department Store«.[16] I have already discussed how the sounds of Big Ben and the specific acoustic properties of a built environment, impart, to ears culturally equipped to recognize it, a time-place specificity to recordings of the *silences* that would otherwise sound mostly like noise. Perhaps what Stephen Feld termed *acoustemology* may also be pursued through silences.[17] Were they able to negotiate the acousmatic dislocation of a recording, the Kaluli people, for example, would perhaps also be able instantly to place the sound of a *silence* in their jungle.

15 Lévi-Strauss 1964, 306-07, 333-36.

16 Schwartz, ibid.

17 Feld's term acoustemology is a compound of acoustic epistemology. It is concerned with the culturally-specific knowledge that might be derived from listening to an acoustic environment, the sounds it contains and the relationship between environmental sounds and cultural soundings (such as the cadences and rhythms of speech and discourse); see Feld 2005 and *passim*.

There are many versions of memorial silence around the world. In the Netherlands, for example, a memorial silence is held at 8pm on 4th May, with its emblematic epicentre in Amsterdam's Dam Square, commemorating the victims of World War II. In Turkey the annual silence on the stroke of 9:05am on 10th November, the moment of Ataturk's death, is topped and tailed by wild noisemaking much as Hillel Schwartz describes in relation to the original UK 1919 silence. Bosphorus ferries, cars and buses all blast horns. The silence itself is then immaculately observed for its two-minute duration, with cars stopping dead in the street, ferries cutting their engines and drifting free, and pedestrians freezing on the spot in impressive synchronicity. But it is the two-minute armistice silence of the 11th hour of the 11th day of the 11th month in Britain that created the template. Let us recap its history.

The idea for a universal *pause* had first been the subject of a popular newspaper campaign originating in a letter to the *London Evening News*, 8 May 1919, from the journalist, Edward George Honey, entitled »Appeal for a memorial silence on armistice day« which called for

> a very sacred intercession, [...] church services, too, if you will, but in the street, the home, the theatre, anywhere, indeed, where Englishmen and their women chance to be, surely in this five minutes of bitter-sweet silence there will be service enough?[18]

At first, this campaign was ignored by a government wary of public protest and reticent to sanction anything that located such obvious power in a locally *ad hoc* and uncontrolled event. This was a time of the post-WWI economic depression, industrial unrest and Spanish flu pandemic – even of mutinous murmurings among the armed forces. Instead, at the beginning of July, the architect Sir Edwin Lutyens was commissioned to design a military parade and ceremony to take place around a cenotaph dedicated to an unknown soldier whose body was to be brought back from France and whose coffin would be paraded up the Mall and down Whitehall to the monument on the 11 November anniversary. However, the popular campaign for the silence grew, and on Nov 4 1919, former South African High Commissioner Sir Percy Fitzpatrick, who had lost his son in France in 1917, wrote to the Colonial Secretary and War Cabinet member, Lord Milner, that:

> During the War, we in South Africa observed what we called the ›Three minutes' pause‹. At noon each day, all work, all talk and all movement were suspended for three minutes that we might concentrate as one in thinking of those – the living

18 Cf. Gregory 2001.

> and the dead – who had pledged and given themselves for all that we believe in [...] Silence, complete and arresting, closed upon the city – the moving, awe inspiring silence of a great Cathedral where the smallest sound must seem a sacrilege [...] Only those who have felt it can understand the overmastering effect in action and reaction of a multitude moved suddenly to one thought and one purpose.[19]

Whether or not this letter directly informed the change of heart, a realization seems to have taken place that an officially-sponsored silence event might have the dual effect of locating the power not only within the paternalistic gift of the State but also within a galvanizing *universal* (i.e. Empire-wide) shared moment – in other words, in a moment of grand theatre. Within days of Sir Percy's letter, the War Cabinet charged Milner with progressing *one* minute's silence following a precedent set at Theodore Roosevelt's state funeral in January that year. In the event, the King, it is believed, increased the allotted time to two minutes, and issued a decree on 7th November 1919 that scheduled the silence for the following Tuesday

> [...] at the hour when the Armistice came into force, the eleventh hour of the eleventh day of the eleventh month, there may be for the brief space of two minutes a complete suspension of our normal activities. No elaborate organisation appears to be required. At a given signal, which can easily be arranged to suit the circumstances of the locality, I believe that we shall gladly interrupt our business and pleasure, whatever it may be and unite in this simple service of Silence and Remembrance.[20]

The two-minute silence also became the centre-piece of Lutyens' parade. The British Movietone Newsreel silent film that disseminated this imagery around the Empire shows the journey of the Unknown Soldier back from France.[21] It then shows the unveiling of the monument, with the moment of the silence framed, *in lieu* of audible chimes and canon blasts, by a shot of Big Ben (in negative so as to provide a black background for an intertitle reading »At the first stroke of the eleventh hour from Big Ben, London was hushed into the Great Silence«). This Great Silence itself is represented by a shot of a busy London street, with motor cars and buses, horse drawn carts, bicycles and milling crowds, cutting abruptly to a shot of the same

19 See National archive.

20 Ibid.

21 See MovieTone Digital Archive.

street-scene standing still, with only the horses' heads moving and the crowd with hats off. The 1923 Pathé news follows the same pattern, with a more prolonged close-up of the clock-face of Big Ben, followed by a shot of the crowd removing their hats.[22] The archival version of the clip holds for a while on the stilled crowd, and then there is a cut to the men putting their hats back on, and a military band and church choir beginning to perform.

This brings us back to where we began, with the silence as a mediatized event. It took until 1929 for John Reith to secure the Home Office permission he had sought since the beginning, to situate BBC microphones at the event. Newsreel footage also acquired sound soon after. As a live event, the circumference of the *silence* now extended via the radio network, across the land and beyond. As well as gathering in local communities to perform the silence around local monuments, or wherever they happened to be at the eleventh hour of the eleventh day of the eleventh month, people now also joined in around parlour radio sets, adopting the now familiar pose. Enacting the *silence* in the home made hyper-real and monumental the sounds of domestic circumstance: of ticking clocks, dripping taps or humming refrigerators. People shared in the greatness and universality of the crowd in their own home rituals, enacted before the altar of the wireless set. Through its acousmatic portal from London came the reverberating, urban noises of Whitehall, illuminating the room and projecting onto the domestic ceremony an imagined scenography of Royals, Generals and politicians standing to attention in the acoustic after-image of Big Ben's chimes. This silence and these urban noises, staged and made epic by a profoundly immersive, intermedial theatre, have a special place within mediated cultural heritage.

22 See British Pathé.

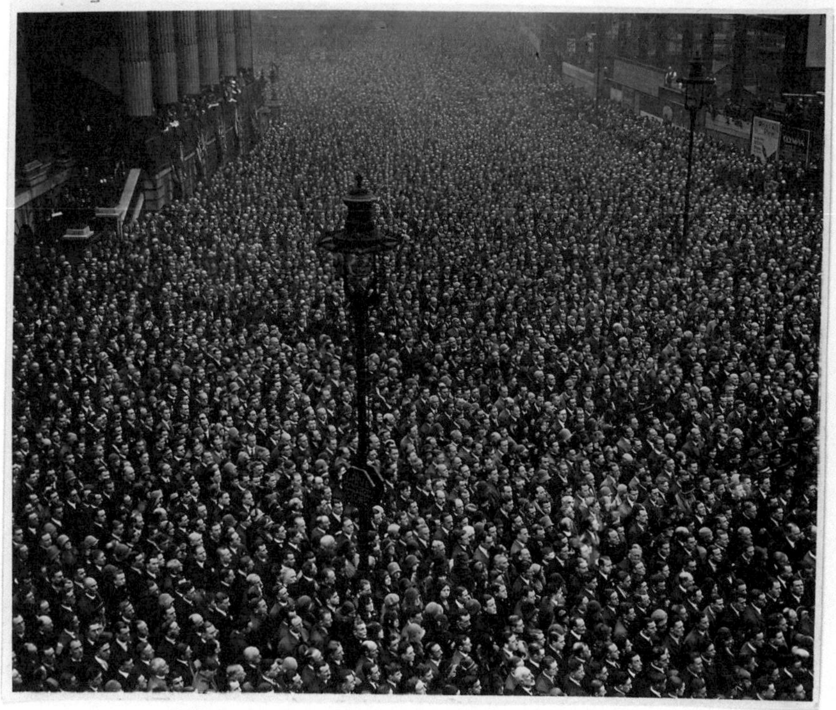

Figure 10: Two-Minute Silence, Armistice Day, London, 1919.

4. Sources

British Pathé: Armistice Day 1923, http://www.britishpathe.com/video/armistice-day-2/query/armistice+day, 2012/04/30.

Brown, Ross (2006): Noise Memory Gesture. The Theatre in a Minute's silence. Arts and Humanities Research Council funded research project, UK

Brown, Ross (2009): Noise, Memory, Gesture. The Theatre in a Minute's Silence, in: Mock, Roberta / Counsell, Colin (eds.) Performance, Embodiment & Cultural Memory, Newcastle, 203-21.

Feld, Steven (2005): Places Sensed, Senses Placed, Towards a Sensuous Epistemology of Environments, in: Howes, David (ed.) Empire of the Senses. The Sensual Culture Reader, Oxford / New York, 179-91.

Gregory, Adrian (2001): The Silence and the History (sleeve note), in: Semper, Jonty (2001) Kenotaphion, Charm Records, UK.

Jackson, Anthony (1968): Sound and Ritual in Man, New Series, 3.2 (June), 293-299.

Lévi-Strauss, C. (1964): Le Cru et le Cuit, Paris.

Maeterlinck, Maurice / Alfred Sutro (translator) (1897): The Treasure of the Humble, London.

MovieTone Digital Archive: story no. 7020, http://www.movietone.com, 2012/04/30.

National archive: ref PRO CAB 24/C45, quoted on http://www.veterans-uk.info/remembrance/ two_minutes.html, 2012/04/02.

Oxford English Dictionary, Third edition (2001): online version March 2012, http://0-www.oed.com.catalogue.ulrls.lon.ac.uk/view/Entry/239856, 2012/05/30.

Schwartz, Hillel (2011): Making Noise. From Babel to the Big Bang and Beyond, New York.

Semper, Jonty (2001i): Kenotaphion, Charm Records, UK.

Semper, Jonty (2001ii): The One-Minute Silence from the Funeral of Diana, Princess of Wales, Charm Records, UK.

The Royal Channel (2010/10/14): Remembrance Day at the Cenotaph 2010 (extract), http://youtu.be/pSol1JxSniU, 2012/05/25.

Authors

Jasper Aalbers, MA

*1984, PhD candidate in the Department of Technology and Society Studies, Maastricht University. His PhD project focuses on the staging and representation of urban soundscapes in fiction film between 1930 and 2000. He is a contributor to the research project Soundscapes of the Urban Past: Staged Sound as Mediated Cultural Heritage (funded by the Netherlands Organization for Scientific Research NWO). He published on the connections between film theory and technology in TMG (Dutch Journal for Media History). Homepage: http://www.unimaas.nl/staff/aalbers

Prof. Dr. Karin Bijsterveld

*1961, Professor in the Department of Technology and Society Studies, Maastricht University. She coordinates several research projects at the crossroads of Science and Technology Studies and Sound Studies (funded by the Netherlands Organisation for Scientific Research NWO). One of these projects is Sonic Skills: Sound and Listening in Science, Technology and Medicine, 1920s-now. She is co-editor (with Trevor Pinch) of the Oxford Handbook of Sound Studies (2012) and a member of the research network Sound in Media Culture (funded by the German Research Foundation DFG). Selected publications: Mechanical Sound. Technology, Culture, and Public Problems of Noise in the Twentieth Century, The MIT Press, Cambridge MA, 2008; Sound Souvenirs. Audio Technologies, Memory and Cultural Practices, Amsterdam University Press, 2009 (co-ed. with José van Dijck).
Homepage: http://www.unimaas.nl/staff/bijsterveld

Dr. Carolyn Birdsall

*1980, Assistant professor in the Department of Media Studies, University of Amsterdam. Her research interests are in the fields of media and cultural history, with a particular focus on radio, film and television sound, documentary and urban studies. Her monograph Nazi Soundscapes (2012) examines the significance of radio and sound systems in urban environments during Weimar and National Socialist Germany. Selected publications: Sonic Mediations. Body, Sound, Technology. Cambridge Scholars Press, 2008 (with Anthony Enns); Earwitnessing. Sound Memories of the Nazi Period, in: Karin

Bijsterveld and José van Dijck (eds.), Sound Souvenirs. Audio Technologies, Memory and Cultural Practices, Amsterdam University Press, 2009.
Homepage: http://home.medewerker.uva.nl/c.j.birdsall/

Prof. Ross Brown

*1963, Dean of Studies and professor at the Central School of Speech and Drama, University of London. He was trained as a fine artist and worked as professional composer, musician, technician, sound designer and writer before joining the academy in 1994. He has played a leading role internationally in establishing sound as a core dramaturgical and scenographic consideration of theatre, as well as a specialist technical art. His research develops discursive frameworks for practice-as-research and theories of theatre sound, noise and aurality that bridge the fields of drama, theatre, sound studies and sonic arts practices. Selected publications: Towards Theatre Noise, in: Lynne Kendrick and David Roesner (eds.), Theatre Noise. The Sound of Performance. Cambridge Scholars, 2012; Sound Design: the Scenography of Engagement and Distraction in: Jane Collins and Andrew Nisbett (eds.), Theatre and Performance Design: a Reader in Scenography, Routledge, London, 2010; Sound. Palgrave Readings in Theatre Practice series, Palgrave Macmillan, Basingstoke, 2009.
Homepage: http://www.cssd.ac.uk/staff/profiles/prof-ross-brown-ba-frsa-fhea

Prof. Dr. Andrew Crisell

*1943, Professor of Broadcasting Studies in the School of Arts, Design and Media, University of Sunderland, United Kingdom. He is Chair of the school's Media Postgraduate Research Group and a member of the editorial boards of the Radio Journal and the Journal of Media Business Studies. Selected publications: A Study of Modern Television. Thinking Inside the Box, Palgrave Macmillan, Basingstoke, 2006; Radio. Critical Concepts in Media and Cultural Studies, (ed.) Routledge, London, 2009; Liveness and Recording in the Media, Palgrave Macmillan, Basingstoke, 2012.
Homepage: http://www.crmcs.sunderland.ac.uk/research-staff/andrew-crisell/

Dr. Andreas Fickers

*1971, Associate professor of comparative media history in the Department of Arts and Culture, Maastricht University. His work focuses on the cultural history of communication technologies and European media history in transnational perspective. With Sonja de Leeuw (Utrecht University) he coordinates the European Television History Network, and he is one of the editors in chief of the Journal of European Television History and Culture. Selected publications: A European Television History (with Jonathan Bignell), Wiley-Blackwell, Hoboken NJ, 2008; Materializing Europe. Transnational Infrastructures

and the Project of Europe (with Alexander Badenoch), Palgrave-Macmillan, Basingstoke, 2010; Transnational Television History (with Catherine Johnson), Routledge, London, 2012.
Homepage: http://www.unimaas.nl/staff/fickers

Drs. Ir. Annelies Jacobs
*1955, PhD-candidate in the Department of Technology and Society Studies, Maastricht University. Her PhD-research focuses on the staging of Amsterdam soundscapes in historical text, which is one of the subprojects in Soundscapes of the Urban Past (a project funded by the Netherlands Organisation for Scientific Research NWO). She has published on the sound of war in Amsterdam in Stadsgeschiedenis (a Dutch journal for urban history, 2012), and on sound as an unheard(of) dimension in architecture in Mosaïk (a Dutch interdisciplinary journal for cultural studies, 2005). Other publications: Storing Sound Souvenirs. The Multi-Sited Domestication of the Tape Recorder (with Karin Bijsterveld), in Bijsterveld, Karin/van Dijck, José (eds.) (2009): Sound Souvenirs. Audio Technologies, Memory and Cultural Practices, Amsterdam.
Homepage: http://www.fdcw.unimaas.nl/staff/jacobs

Evi Karathanasopoulou
*1980, PhD candidate at the University of Sunderland, United Kingdom. Her area of research is Radio Studies with a focus on aesthetics, radio art and the radio voice. She has earned a BA (First Class Hons) in Media Production (Television and Radio) from the University of Sunderland and an MA in Radio from Goldsmiths College, University of London. In February 2008 she received an Award for excellence for the best Short-form Radio Report from the Broadcast Journalism Training Council (BJTC) in the United Kingdom. She co-authored Teaching Sustainable Media Practice (with Alex Lockwood and Caroline Mitchell), in: Kristen Allen Bartels and Kelly A. Parker (eds.), Teaching Sustainability, Stylus, Sterling VA, 2012.
Homepage:http://www.crmcs.sunderland.ac.uk/postgrad-research/current-postgrad/evi-karathansopoulou/

Prof. Dr. Patricia Pisters
*1965, Professor of media culture and film studies, and chair of the Department of Media Studies, University of Amsterdam. She has published on film-philosophical questions of audio-visually mediated perception, the ontology of the image, the politics of contemporary screen culture, transnational media, and the idea of the brain as screen in connection to neuroscience. Selected publications: The Matrix of Visual Culture. Working with Deleuze in Film Theory, Stanford University Press, 2003; Shooting the Family.

Transnational Media and Intercultural Values (co-ed. with Wim Staat), Amsterdam University Press, 2005; Mind the Screen (co-ed. with Jaap Kooijman and Wanda Strauven), Amsterdam University Press, 2008; The Neuro-Image. A Deleuzian Film-Philosophy of Digital Screen Culture, Stanford University Press, 2012.
See also www.patriciapisters.com

Prof. Dr. Holger Schulze
*1970, Visiting professor for Sound Studies at the Humboldt-Universität Berlin. Principal investigator of the Sound Studies Lab and director of the international research network Sound in Media Culture (both funded by the German Research Foundation DFG). Founding editor of the Sound Studies book series. Selected publications: Intimität und Medialität. Tektonik der Medien, Berlin 2012; Sound Studies. Eine Einführung, (ed.) Transcript, Bielefeld 2008; Klanganthropologie. Performativität – Imagination – Narration, (co-ed. with Christoph Wulf) Akademie Verlag, Berlin 2007; Heuristik. Theorie der intentionalenWerkgenese, Transcript, Bielefeld 2005.
Homepage: http://www.soundstudieslab.org/author/admin/

Prof. Mark M. Smith
*1968, Carolina Distinguished Professor of History, University of South Carolina, USA. He is a scholar of US history and the history of the senses and author or editor of a dozen books. His work has been translated into Chinese and Korean and has been featured in the London Times and the New York Times. He is a former president of The Historical Society. Selected publications: Camille, 1969. Histories of a Hurricane, University of Georgia Press, Athens GA, 2011; Sensing the Past, Seeing, Hearing, Smelling, Tasting and Touching in History, University of California Press, Chapel Hill NC, 2008; How Race Is Made. Slavery, Segregation, and the Senses. University of North Carolina Press, Chapel Hill NC, 2006.
Homepage: http://www.cas.sc.edu/hist/facultyprofiles/smithmark.html

Prof. Dr. Jonathan Sterne
*1970, Associate professor in the Department of Art History and Communication Studies, McGill University, Canada, with additional affiliations with the History and Philosophy of Science Program, the Institute for Gender, Sexuality and Feminist Studies, and the Centre for Studies in Improvisation. He has written widely on sound and music, the history and philosophy of technology, cultural studies and new media. He is editor of The Sound Studies Reader, Routledge, London, 2012. Current projects include a study of the aesthetics and politics of signal processing (funded by the Social Sciences and Humanities Research Council of Canada), and a study of media and the politics of inhabitability

(as part of a team funded by the *Fonds Québécois sur la Société et la Culture*). He is also a member of the Sound in Media Culture research network (funded by the German Research Foundation, DFG), and Centre de recherché sur l'intermédialité. Selected publications: The Audible Past. Cultural Origins of Sound Reproduction. Duke University Press, Durham NC, 2003; MP3. The Meaning of a Format, Duke University Press, Durham NC, 2012.
For more information, see his website at http://sterneworks.org

List of Illustrations